The radiotoday

guide to the

ICOM IC-7300

by

Andrew Barron, ZL3DW

Radio Today is an imprint of the Radio Society of Great Britain

This book has been published by Radio Society of Great Britain of 3 Abbey Court, Priory Business Park, Bedford MK44 3WH, United Kingdom
www.rsgb.org.uk

First Edition 2019

This Revised Edition first printed 2021

Reprinted digitally 2022 onwards

The opinions expressed in this book are those of the author and are not necessarily those of the Radio Society of Great Britain. Whilst the information presented is believed to be correct, the publishers and their agents cannot accept responsibility for consequences arising from any inaccuracies or omissions.

The author has no association with Icom, any Icom reseller, or service centres. The book is not authorised or endorsed by Icom or by any authorised Icom Dealer or repair centre. Research material for the creation of this document has been sourced from a variety of public domain Internet sites and information published by Icom including the Basic Manual and the Advanced Manual.

ISBN: 9781 9101 9373 0

Cover design: Kevin Williams, M6CYB
Typography and design: Andrew Barron ZL3DW
Production: Mark Allgar, M1MPA

Printed in Great Britain by 4Edge Ltd. of Hockley, Essex

Any amendments or updates to this book can be found at:
www.rsgb.org/booksextra

The Radio Today guide to the Icom IC-7300

By Andrew Barron ZL3DW

Table of Contents

OTHER BOOKS BY ANDREW BARRON

The Radio Today guide to the Icom IC-7610

The Radio Today guide to the Icom IC-705

The Radio Today guide to the Icom IC-9700

The Radio Today guide to the Yaesu FTDX101

Testing 123
Measuring amateur radio performance on a budget

Software Defined Radio
for Amateur Radio operators and Shortwave Listeners

Amsats and Hamsats
Amateur radio and other small satellites

An introduction to HF Software Defined Radio
(out of print)

ACKNOWLEDGEMENTS

Thanks to my wife Carol for her love and support and to my sons James and Alexander for their support and their insight into this modern world. Thanks also to Icom who produced the excellent IC-7300 transceiver and finally, many thanks to you, for buying my book.

DEDICATION

This book is dedicated to the families, friends, and loved ones of the 50 people killed and 92 people injured by a racist extremist on Friday 15th March 2019 in my home town of Christchurch, New Zealand. We view these atrocities as happening in another country, or another town, or to a group of people who are in some way different to ourselves. How close to home does violence have to come before we realise that the victims of these many outrages are, our people, from our world? Just ordinary people like us. I ask you to please remember that the next time you hear someone tell a racist joke or treat someone badly because of their religious faith or the colour of their skin. Racism and religious intolerance breed violence.

ACRONYMS

Amateur radio is chock full of commonly used acronyms and TLAs (three letter abbreviations :-) They can be very confusing and frustrating for newcomers. I have tried to expand out acronyms and explain abbreviations the first time that they are used. Near the end of the book, I have included a comprehensive glossary, explaining the meaning of many terms used in the book. My apologies if I have missed any.

The Icom IC-7300 transceiver

Congratulations on buying or being about to purchase the amazing Icom IC-7300. It is always exciting unboxing and learning how to use a new transceiver. The IC-7300 has created something of a revolution in the amateur radio world. With this radio, Icom provides the advantages of SDR technology in a format that is familiar for users of their earlier transceivers. Most importantly the IC-7300 has many features that were previously only available on much more expensive radios.

This revised edition of the Radio Today guide to the Icom IC-7300 includes the new features added in the February 2021 v1.40 firmware update. These include the very useful FT8 Preset, which can be used for all external digital modes, scrolling the spectrum display if you tune off the edge of the currently displayed span, four fixed edges for the FIX display, using the MULTI control to step through memory channels, the ability to temporarily allocate one of fifteen transceiver adjustments to the MULTI knob, and the ability to apply custom functions to four of the front panel buttons, and two microphone buttons.

If this is your first SDR I am sure that you will be excited about the panadapter. I particularly like the 'FIX' spectrum display mode where you can see the whole band or just a section of the band such as the CW segment. This has been expanded from three to four saved FIX spectrum edges per band.

Compared to other transceivers there are quite a few changes to the way you operate the radio. These are mostly due to the touch screen controls. As I have got to know the radio by using it and through delving into every control and menu setting, I have discovered many clever design features. You can certainly see the benefit of decades of Icom technical development and experience.

The IC-7300 has a fairly simple front panel layout, with many settings being accessed through the touch screen display. For example, there are no physical 'band change' or 'mode' buttons. Less often changed settings are available through the MAIN, FUNCTION, and QUICK menu buttons. It can be a bit of a challenge remembering which settings are on what sub-menu and that's what this book is all about. There are step by step instructions for setting up the radio and I have included a 'quick reference guide' at the back so that you can easily find your way through the menus to make the changes that you want.

The IC-7300 is a good radio for chasing DX, 'rag chewing,' or contesting. I enjoy alternating between the fixed spectrum scope which shows you all the signals across the band, and the 'Centre' display which shows you a few signals either side of the VFO centre frequency.

I love the SSB voice message keyer. It makes calling "CQ" or just saying my callsign repeatedly while trying to work a 'DXpedition,' so much easier.

You can operate RTTY without relying on a connection to a 'digital modes' program running on your computer. There are eight pre-set messages and the RTTY decoder works quite well. Sadly, you can't send directly from an external keyboard.

The inclusion of the 'panadapter' or 'band-scope' marks out the radio as being excellent value for its price point. But there is a lot more. The radio has VOX (voice operated transmit switching), a built-in antenna tuner, coverage of the 6m (50 MHz) and 4m (70 MHz) amateur radio bands. And it has a voice keyer, RTTY and CW keyer memories, and an RTTY decoder.

In fact, the radio has virtually every option that you could want. It combines cutting edge SDR technology with all the controls and features that an experienced amateur radio operator expects from an Icom transceiver. Of course, that makes setting it up quite a complicated task. There are many opportunities to configure the radio to your tastes. The configuration settings are available through the touch screen menus. It can be a challenge remembering which menu sets what function. I found that I was changing a lot of settings as I customised the radio to my liking. I began to lose track of the default settings, although they are all in the manual. I decided to write down the way that I have the radio configured for future reference. You will probably decide on your own preferences. I left room for you to add your preferred settings to the relevant tables in the 'Menu' section.

Coming from several years using a "black box - no knobs" Apache Labs ANAN-100 SDR transceiver, I am a bit disappointed by the inability to increase the gain of the spectrum display. You can set the reference level, but you can't make the spectrum display more sensitive. The IC-7610 and IC-9700 transceivers have the same problem.

If you use external software for digital modes such as PSK or FT8, I am sure that you will find the section on configuring the USB CI-V interface between the radio and your PC very helpful.

I cover some "weird" operating behaviours in the troubleshooting chapter. If you experience something strange, have a look to see if it is covered in that chapter. I am sure that some of these oddities will be addressed in future firmware releases. Finally, there is a chapter listing some of the modifications that some users are carrying out. Please note that I am not endorsing any modification to the radio and that carrying out any modification will almost certainly void the Icom warranty. Last of all there is an alphabetical list of the menu commands require to perform most of the setup options.

TECHNICAL FEATURES

The IC-7300 is an SDR (software defined radio) or more correctly a direct digital sampling radio. If you have not used an SDR transceiver before, you will be impressed by how clean the receiver sounds and you will quickly get used to the advantages of the panadapter display. Icom has kept the layout very similar to their previous radios and anyone transitioning from an IC-7410 or even an IC-746 pro will be very comfortable. One or two controls are available as both conventional buttons and Soft Key functions. For example, you can turn on the preamplifiers using the P.AMP button or by pressing FUNCTION and using the P.AMP Soft Key.

The screen is a 4.3" (93mm x 52mm) colour TFT LCD display. It has a very clear, crisp image, with excellent contrast and colour saturation. There is an adjustable LCD backlight to ensure that the brightness will suit indoor or outdoor operation.

The spectrum scope and waterfall display are significantly better than the ones on earlier radios although the small screen can't compete with black box SDRs that display their panadapters on a computer monitor.

The spectrum scope has a maximum bandwidth of one megahertz, which is the same as the panadapter on the IC-7610 and twice the bandwidth of the scope in the IC-7600. It is also faster and has a higher refresh rate. Before I bought the radio, I wondered if a 1 MHz span would be enough. But it is more than enough. Only the ten and six metre bands are wider than a megahertz and you can set up the spectrum scope band edges to display those bands in sections.

Direct digital sampling is used for both the receive and transmit signal paths. The receiver chain starts with an RF preselector stage containing fifteen bandpass filters, or a low pass filter for frequencies below 1.6 MHz. These are very effective at keeping 'out of band' signals out of the receiver. The next stage is the ADC (analogue to digital converter) followed by the FPGA (field programmable gate array) which performs digital down conversion to a 12 kHz wide digital I.F. centred at 36 kHz. This is followed by a conventional Texas Instruments TMS320C6745 DSP (digital signal processing) stage which provides the filtering and demodulation. After that, the signal is converted back into an analogue audio signal and sent to an audio amplifier to drive the speaker and the headphone jack. There is a separate DSP stage for the panadapter display. The transmitter uses an ADC to convert analogue signals from the microphone or other analogue input into a digital signal. This is followed by a DSP and digital up-conversion in the FPGA. After that the digital signal, which is already carrying the transmission at the wanted transmit frequency is converted to an analogue signal using a DAC (digital to analogue converter). This is followed by the same bandpass filters used in the receiver chain, the ALC (automatic level control) stage, and the RF power amplifier which boosts the signal up to the 100W level. Finally, there is a bank of seven low pass filters for harmonic suppression, the same as you would see in any transceiver.

Icom is justly proud of the 100 dB RMDR measurement (2 kHz spaced reciprocal mixing dynamic range) achieved by the receiver. This test result is better than the RMDR results achieved by the earlier IC-7200 and IC-7100 transceivers. The receiver two tone IMD results are average for a recent transceiver, placing the radio at the 18th spot on the Sherwood Engineering list. The transmitter has exceptional transmit 3rd order IMD results, with low transmitted phase noise and harmonics.

The IP+ mode is designed to improve the receiver's intermodulation performance by turning on the ADC 'dither' and 'randomisation' functions. This does come at the expense of a small reduction in the receiver sensitivity. It improves the lab test results, but it is usually unnecessary in real-world conditions.

Some online commentators say that they don't like touch screens. They find them fiddly and complain of fingerprints. This is NOT a radio for them. It is impossible to use the radio without touching the screen. However, I find that unless you've been eating jam sandwiches the display does not show fingerprints. While one or two touch screen functions are duplicated with "real" buttons, all the menu structure and even basic functions like changing the mode or the band can only be changed via the touch screen display. Anyway, I really like it. It works very well, and it is crisp, bright, and colourful with very high definition.

Other notable features that I really like, include;

- The built-in antenna tuner.
- The VFO knob, which is smooth and nicely weighted. You can adjust the drag, but I am happy with it the way it was supplied from the factory.
- A great squelch system – so I don't have to listen to the band noise
- Auto-Tune to pull CW signals to the correct tone
- The built-in RTTY mode with a message keyer and on-screen decoder
- The voice keyer with its eight voice messages. Brilliant!
- The FIX spectrum display mode. You can set the band edges to any band segment that is less than 1 MHz wide. There are four band edges for each of thirteen frequency ranges. For example, on the 20m band, I have one set up to show the whole 20m band, one for the CW band segment, and the third displays the digital mode part of the band.
- The recording mode which lets you record signals off air and optionally your transmissions as well.

Approach

Rather than duplicate the manuals which describe each button and control, I have used a more functional approach. This is a "how to" book. For example, I describe how to set up the transceiver for SSB operation, including all the relevant menu settings. Then I follow that up with setting up instructions for CW, FM, RTTY, PSK, and external digital mode software such as FT8.

Although the IC-7300 has a simple front panel and is easy to operate, the touch screen controls will be new to many and there are a lot of configuration options. Especially if you want to use external digital mode software such as WSJT-X, Fldigi, MixW, CW Skimmer, or MRP40 to mention a few. I was quite surprised at the number of things that had to be configured. Icom has allowed you to make things exactly the way you want them. I imagine that you will experiment with some of the settings more than once before you decide on the optimum settings for your radio.

There are sections on updating the radio firmware, loading the Windows driver software for the USB cable connection, connecting the radio to your PC for CI-V control, and setting the clock. I even cover linear amplifier connections and the FM mode. How to set up tones for repeater access and store repeater channel frequencies in the radio memory slots.

The 'Setting up the radio' chapter is followed by information about the front panel and touch screen controls and the MULTI menus. Then 'Operating the radio' in various modes including Split operation. After that comes information about every MENU item and FUNCTION setting.

The 'Useful Tips' section describes the screen saver and a special setting for the spectrum display.

The Troubleshooting section deals with some oddities that might trip you up.

The Modifications setting covers some of the modifications that people applying to the radio. Please note that I am not endorsing any modification to the radio and that carrying out any modification will void the Icom warranty.

The Glossary explains the meaning of the acronyms and abbreviations used throughout the book and the Index is a great way of going directly to the information that you are looking for.

Last, of all, there is the Quick reference guide, where you can look up the menu steps for most of the commonly used functions. I find it useful because I can never remember where to find the ones I want.

Conventions

The following conventions are used throughout the book.

Front panel controls and buttons are indicated with a highlight. `TRANSMIT`

Touch screen controls are indicated in uppercase without a highlight. AGC MID

'Touch' means to briefly and gently touch the item on the touch screen. There is no need to press hard on the screen.

'Touch and hold,' means to keep touching the item on the touch screen for one second or until the function changes. It usually opens a control option or sub-menu window on the touch screen.

'Press' means to press a physical button or control.

'Press and hold' or 'hold down' means to hold a physical button or control down for one second or until the function changes. It usually opens a control option or sub-menu window on the touch screen.

If 'Beep' is set to on, all touch and press operations are accompanied by a quiet beep. Hold operations are often accompanied by two beeps.

Words in <brackets> indicate a step in a sequence of commands. The sequence usually begins with pressing a <button> followed by touching a series of <Soft Keys> or <icons> on the touch screen.

A 'Soft Key' is an icon on the touch screen that represents a button or control.

The 'Return' or 'Exit' Soft Key is a back-turning arrow. Something like this ↰.

Mode usually means one of the transceiver modulation modes. SSB, CW, RTTY, AM, FM, and Data (USB-D, LSB-D, AM-D, or FM-D). But it can also mean 'CENT' band scope mode, 'FIX' panadapter mode or an external digital mode performed by software running on your PC.

'Panadapter' or 'Spectrum Scope' means the spectrum and waterfall display. The radio can display either a 'Band Scope' in which the main VFO is always in the centre and the scope displays the spectrum below and above the VFO frequency. Or a 'Panadapter' in which the receiver can be tuned to a frequency anywhere across the range of frequencies that are displayed. To avoid confusion, I use the term 'panadapter' whether the display is in 'CENT' band scope mode, or in 'FIX' panadapter mode.

Setting up the radio

The Icom manuals do a good job of identifying all the controls and menu items, but they lack 'step by step' instructions on how to set each control correctly. In this chapter, I cover the processes that you need to follow to get ready for operating the radio.

Each section lists the strings of menu commands that you need to follow when you set the audio levels, tone settings and transmitter bandwidth. It is important to set the Mic Gain and Compression controls correctly for SSB operation. I also explain the setup for the FM, CW, and RTTY modes including the built-in options and using external digital mode programs. There are instructions on configuring the eight voice messages for SSB and the pre-defined keying messages for CW and RTTY.

I tell you how to download and install the USB driver software to allow the radio to communicate with your computer and discuss setting up the virtual COM ports and the Audio CODEC, linear amplifier connections, and the increasingly popular FT8 mode. Finally, I cover formatting and using the SD card, how to do firmware updates, and how to set the clock.

The radio is very easy to configure once you know which menu settings to use. Where I have changed my radio from the Icom default settings, I explain why I made the change and the effect that it has on the radio. The big advantage of using the instructions in this section rather than using the Icom manual is that I have included all the necessary steps, and the optional ones, in one place. I tell you what the controls and settings do and how they should be adjusted.

DISPLAY YOUR CALLSIGN WHEN THE RADIO STARTS

It is nice to personalise the radio by having it display your name or callsign as the radio 'boots up.' You can select exactly what is displayed upon start-up.

Select <MENU> <SET> <Display>.

- <Opening Message> selects if you want to see the Icom IC-7300 screen on start-up. "Of course, you do!"

- <My Call> allows you to enter your callsign or your name (up to 10 characters).

- Setting <Power ON Check> to ON displays the current RF Power setting on start-up.

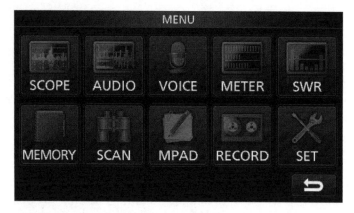

The MENU screen in SSB mode. FT8 Preset on <2>

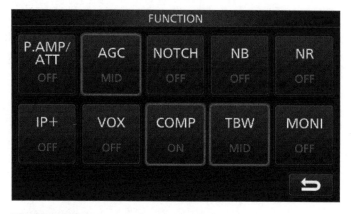

The FUNCTION screen in SSB mode.

The touch screen Soft Keys

SETTING UP THE RADIO FOR SSB OPERATION

There are two settings that should be adjusted before transmitting on SSB. You need to set the Mic Gain control correctly and the Compression level control. You don't have to use the compressor if you are primarily interested in Rag Chewing, but you might as well set it anyway. The instructions in the manuals are a bit vague, but the process is simple, and you only need to do it once. The default levels turned out to be fine for my voice using the supplied hand microphone. The good news is that the radio software will not allow you to exceed 100 Watts of output power and the ALC (automatic level control) limits the audio. It won't allow the microphone audio to overload and distort the modulation. So even if you leave the Mic Gain at the default (50%) setting, the radio will probably sound OK. That said, it is better not to overdrive the system.

In the absence of a clear guide, I have devised my own method for setting the levels. This may not be how Icom or other commentators do it, but it works for me.

1. First, we will set the MIC GAIN control on SSB mode with the Compressor turned off.

    a. With the radio connected to a 50 Ohm dummy load or a tuned antenna on a dead band. Set the radio for the correct mode USB (upper sideband) or LSB (lower sideband) on a frequency in the 'Phone' part of the band that you use the most. If in doubt use USB on the 20m band at about 14.2 MHz. Make sure that you have not selected a data mode. If the blue icon says USB-D or LSB-D touch DATA to set it back to a voice mode.

    b. Press the MULTI knob and touch the RF POWER Soft Key to turn the icon blue. Turn the Multi knob to set the RF POWER to 100%.

    c. If the compressor is turned on, touch the COMP Soft Key on the MULTI sub-menu to turn the compressor off (COMP OFF). Exit the MULTI menu.

    d. I like to see all the metering at once by selecting <MENU> <METER>. But if you like you can just select the relevant meter reading by touching the meter scale. Repeatedly touching the meter scale cycles through Po (power out), SWR, ALC, COMP, V_D, and I_D. Press M.SCOPE to get rid of the panadapter display.

    e. Press MULTI to show the MULTI menu and touch the MIC GAIN Soft Key to activate the control.

    f. Hold the microphone and press the PTT button. Speak into the microphone in the same way that you would while making a contact or calling CQ. Watch the ALC reading on the Multi-Function meter or if you are not using the multi-function meter select ALC on the meter.

While speaking, adjust the MIC GAIN while observing the ALC reading. The ALC meter reading should remain within the red marked zone occasionally peaking to near the top of the zone. But it should **never** peak above 50%. It should always remain within the area marked in red. Running the ALC over 50% will cause your transmitted signal to sound distorted. It will not give you more output power. The RF Power output should peak to 100%.

It is good practice to use your callsign and that tell the world that you are "testing," especially if you are transmitting into your antenna.

    g.   For my station at home, I ended up leaving the MIC GAIN at the default setting of 50%. Your setting may be higher if you speak more quietly than I do.

2.   Next, with the compressor turned on, we will set the Compressor level.

    a.   If necessary, press MULTI to show the MULTI menu. Then touch the COMP button to turn on the compressor. A blue indicator beside the Soft Key indicates that compression is on. The text should now say COMP ON. You also get a COMP indicator just below the VFO MHz display.

    b.   I like to see all the metering at once by selecting <MENU> <METER>. But you can just select the relevant meter reading by touching the meter scale. You will have to check the ALC and COMP readings as you speak.

    c.   Hold the microphone and press the PTT button. Speak into the microphone in the same way that did while setting the MIC GAIN. It is important that you don't shout. You should speak as you would while on the air. Set the COMP LEVEL control so that the COMP meter ranges in the middle of the zone between 10 and 20 but never peaking over 20. I prefer my voice to sound less compressed, so I set the COMP LEVEL to the default setting of 5 which makes the compression a bit lower. On the multi-function meter, I get regular compression up to a reading of 10 with occasional peaks up to 20.

    d.   That's all you need to do for SSB transmission.

3.   One of the really neat features of the IC-7300 is the voice message keyer macros. If you plan to use the voice message keyer, you will need to set the output level of that so that it modulates the transmitter to the same level as the microphone.

You want listeners to be unable to detect when they are listening to a recorded message.

Before setting up the levels you will have set up some voice keyer messages, (see page 95). Once that has been completed, return to this page to set the levels.

To set the VOICE TX voice keyer transmit level.

a. Select the screen that has the larger edge meter and no panadapter.

b. Bring up the VOICE TX screen by pressing <MENU> <VOICE>

c. Touch the TX LEVEL Icon and then trigger one of the voice messages. Use the MAIN VFO knob to set the transmit level. Set the level so that you get peaks to 100% power. Try sending the same message by talking into the microphone. The aim is to have the recording sound exactly like you are using the microphone. I ended up leaving the TX LEVEL at 50%.

d. You can check that the level is about right by selecting the ALC meter by touching the meter scale until the right meter appears. The ALC reading is similar when keying a voice message to when you are speaking into the microphone. Similarly, if you are using the compressor, you can select the COMP meter and compare the amount of compression being developed. Reduce or increase the TX LEVEL until the readings are similar or a little less than the microphone levels. Do not adjust the COMP level or the MIC GAIN or you will end up going in circles.

e. If you have set the Auto Monitor setting to ON (default) you will be able to hear the message as it is transmitted. <MENU> <VOICE> <REC/SET> <SET> <Auto Monitor> <ON>.

f. Touch and holding any voice keyer 'M' message icon causes the message to repeat at intervals set by the Repeat Time function. <MENU> <VOICE> <REC/SET> <SET> <Repeat Time>.

SETTING THE TRANSMIT BANDWIDTH (TBW)

If you are in SSB mode, you can select from a choice of three transmit audio bandwidths. Press the FUNCTION button and then touch the TBW Soft Key to cycle through NAR, MID, or WIDE.

WIDE is suitable for Rag Chewing on the local Net or chatting to locals on the 80m or 10m band. MID is better suited to working DX and the NAR (narrow) mode is suited to contest operation.

If you are happy to use the default settings, that is all you need to know. However, you can set the transmit bandwidth for each of the three options if you want to.

➤ **SSB tone controls and transmitter bandwidth**

- Select <MENU> <SET> <Tone Control/TBW> <TX> <SSB>

- You can adjust the Bass and Treble of your transmitted audio. I don't recommend making any big changes, but I looked at the transmit audio spectrum while transmitting and adjusted the treble to +2 to make my transmitted audio spectrum a little flatter. <MENU> <AUDIO> opens the audio scope screen.

- The next three menu items are for adjusting the transmit bandwidth on Wide, Mid and Narrow.

    o Wide default is 100 Hz to 2900 Hz

    o Mid default is 300 Hz to 2700 Hz. (I changed mine to 200 – 2700 Hz)

    o Nar default is 500 Hz to 2500 Hz. (I changed mine to 300 – 2500 Hz)

- Touch and hold any TBW setting to reset the bandwidth to the Icom default.

➤ **SSB-D data mode transmit bandwidth**

- Select <MENU> <SET> <Tone Control/TBW> <TX> <SSB-D>

The default transmitter bandwidth for the SSB data modes is 2.4 kHz (300 Hz to 2700 Hz). This is less than the 3 kHz (widest) receive filter for the SSB-D data mode. Which means that your external digital mode program will display a wider panadapter than you can use for transmitting. If this bothers you, you could reduce the bandwidth of the USB-D FIL1 receiver filter to 2.4 kHz, then they will match. Generally, you will use the middle or narrow receive filter for external digital modes in which case it is not an issue. I decided to leave the TBW at default and to reduce the USB-D FIL1 filter bandwidth to 2.8 kHz. Halfway between the original 3 kHz receiver filter and the 2.4 kHz transmit bandwidth.

The maximum frequency you can set is 2900 Hz and the minimum is 100 Hz (2.8 kHz bandwidth). Touch and hold the TBW setting to reset the bandwidth to the Icom default. 300 Hz to 2700 Hz (2.4 kHz bandwidth).

Note: if you are planning to use the radio for the FT8 mode, the transmit bandwidth doesn't matter because the WSJT-X program will use the Split mode to offset the transmit signal. So, there is no need to push out the transmit bandwidth.

➤ **AM and FM modes transmit bandwidth**

- Select <MENU> <SET> <Tone Control/TBW> <TX> (<AM> or <FM>)

You can't adjust the transmit bandwidth for the AM or FM modes, but you can adjust the transmitted bass and treble.

SETTING THE RECEIVER TONE CONTROLS

➢ **RX HPF/LPF adjustment**

You can set audio high pass and low pass filters for all modes except the DATA modes. Select <MENU> <SET> <Tone Control/TBW> <RX> <*choose mode*> <RX HPF/LPF>. I have not made any changes as the default settings seem OK to me. Note that adjusting the RX HPF/LPF will override the bass and treble setting and return those controls to zero.

➢ **Bass and Treble adjustment**

You can adjust the Treble and Bass for the three voice modes, SSB, AM, and FM. Select <MENU> <SET> <Tone Control/TBW> <RX> <*choose mode*> <RX Bass> or <RX Treble>. Note that if you change the bass or treble setting, it will override the RX HPF/LPF setting and return it to default '- - - - - - -.'

SETTING UP THE RADIO FOR CW OPERATION

There is no real setup required for the CW mode. No levels to set other than the transmitter power and the sidetone level. Note that the MONI (transmit monitor) function is disabled in CW mode because the sidetone is always turned on. If you don't want sidetone you can turn down the level to zero. (See note below).

➢ **Keys**

There are provisions for different types of 'Morse' key. The rear panel 'Key' jack can be used with a Paddle, Bug, or Straight Key. You can even send CW using the up and down buttons on the Icom hand microphone. The wiring is standard with 'dots' on the phono plug tip, 'dashes' on the ring, and common on the sleeve. See page 131 for more information.

➢ **CW settings**

The CW controls are only accessible when the radio is in the CW mode. The most often used controls for CW operation are the key speed and pitch. These can be adjusted at any time, to suit your needs. Press the MULTI button and select KEY SPEED or CW PITCH. Turning the Multi Knob adjusts the selected item. The CW speed is adjustable from 6 wpm to 48 wpm. The CW pitch is adjustable from 300 Hz to 900 Hz. I use 700 Hz.

The other menu settings for CW are on the KEYER menu. Set the radio to CW mode, then <MENU> <KEYER> <EDIT/SET> <CW-KEY SET>. There are eight menu options spread over two screens. You probably won't need to change any of the options. But there is more about this on page 99.

➢ **Break-in setting**

Your CW (Morse Code) signal won't automatically be transmitted unless either full break-in or semi break-in has been selected. When the radio is in CW mode, press the FUNCTION button and choose either BKIN or F-BKIN and the radio will transmit when you operate the key or trigger a message macro.

If you want to practice your CW without transmitting, you can listen to the sidetone when break-in is set to OFF. To transmit in that mode, you have to manually key the transmitter by pressing the TRANSMIT button, pressing the PTT button on the microphone, using a CI-V command, or by grounding the SEND line on the ACC jack. By the way, you cannot key the transmitter by grounding the SEND RCA jack on the rear panel.

The break-in settings affect the sending of keying macro messages as well as CW sent from a key or paddle. But they have no effect on CW sent from a PC application.

- With BK-IN OFF you can practice CW by listening to the side-tone without transmitting.

- F-BKIN. The full break-in mode will key the transmitter while CW is being sent and will return to receive as soon as the key is released. This 'QSK' mode allows for reception of a signal between CW characters.

- BKIN. The semi break-in mode will key the transmitter while CW is being sent and will return to receive after a delay. Touch and hold the F-BKIN or BKIN Soft Key to adjust the delay. Turn the Multi knob to change the setting. The default is a period of 7.5 dits at the selected CW keying speed.

➢ **Sidetone**

CW sidetone is 'always on' but you can set the level to zero if it is annoying, or if your key generates its own sidetone. In CW mode select <MENU> <KEYER> <EDIT/SET> <CW-KEY SET> <Sidetone Level>.

➢ **CW message keyer**

When in CW mode press MENU then KEYER to show the eight CW messages. They can be used for DX or Contest operation or just to save you sending the same message over and over. They are great for sending CQ on a quiet band.

Touching one of the **M1 to M8 Soft Keys** sends the CW message. Touch again or send a dit or dah from the paddle to stop sending the message.

Touch and hold one of the **M1 to M8 Soft Keys** to keep sending the message until you stop it by touching the Soft Key again or by sending with the key or paddle.

At a minimum, you will have to add your callsign or else the macros will send "Icom" in place of it.

To edit the messages. In CW mode select <MENU> <KEYER> <EDIT/SET> <EDIT>. Touch and hold the message that you want to edit. Then select the EDIT option.

An onscreen keyboard will appear, and you can edit or replace the message. Make sure that you press ENT, or your changes will be lost when you exit the keyboard screen.

> **Contest number CW mode**

In the M2 message, you can enter a star (*) after the signal report and the radio will update the contest number used for contest reports automatically. This auto numbering star can be used in any one of the eight macros and by default it is included in macro M2. If the star has been used in any other macro, it is not shown on the list of available symbols when you edit any of the others. An upward arrow beside the 'M2' indicates that the M2 macro is the one with auto numbering. If you want to shift the function to another macro, you have to remove the star from M2 first. You also have to change the 'Count Up Trigger' setting to the new macro. It's easier just to leave it on the M2 macro.

If you don't work contests, you can leave out the star and change the M2 message to something you can use.

A Caret ^ symbol removes the space between two letters, for example, ^AR or ^BK.

On the <MENU> <KEYER> <EDIT/SET> <001 SET> menu you can set

    a) The number style used for automatic contest numbers.

        a. Normal numbers 001 etc.

        b. ANO style A=1, N=9, O=0

        c. ANT style A=1, N=9, T=0

        d. NO style N=9, O=0

        e. NT style N=9, T=0

    b) The macro using the automatic 'Count Up' feature? The default is M2. It must be the macro that has the star (*) in the text.

    c) The start number, usually 001. Note that if you get a busted contest QSO you can decrement the counter by touching -1 on the Keyer message screen.

Touch the Return icon ↻ or the EXIT button to exit each menu layer.

➢ **¼ tuning speed**

The digital modes and CW allow the use of the ¼ tuning function. Select <FUNCTION> <1/4> to turn the function on or off. The function slows down the tuning rate of the VFO to make tuning in narrow CW and digital mode signals easier. It is indicated with a ¼ icon to the right of the 10 Hz digit of the frequency display.

SETTING UP A CONNECTION TO YOUR PC

The USB port on the rear panel of the radio is used for CI-V CAT control of the transceiver and for transferring audio to and from a PC for external digital mode software. It is a USB 2.0 Type B port. You need a Type A to Type B USB 2.0 cable to connect the radio to a PC. If you have a spare USB 2.0 port on the PC, feel free to use it as there is no advantage in wasting a USB 3.0 port.

➢ **Driver software**

To use the USB port, it is essential that you download and install the Icom driver software on to your PC. The Icom driver software can be downloaded from the Icom website. http://www.icom.co.jp/world/support/download/firm/index.html.

Make sure that you download the version that is suitable for the IC-7300. The USB driver is not in the IC-7300 section, it is further down the list. The current version is USB Driver (Version 1.30).

IC-7100/ IC-7200/ IC-7300/ IC-7410/ IC-7600/ IC-7610/	USB Driver(Version 1.30), Driver Utility and manuals.	2018/06/07

The driver software creates a virtual COM port. On my PC it is COM8. The port that the driver software selects depends on what is already in use on your computer. You will need to know the COM port number when you set up the digital mode or other PC software. The COM port is used for CI-V (CAT) control of the radio from the PC. Its RTS and DTR lines are used to key the radio to transmit and to send CW or digital mode data.

In Windows 10 select 'Settings' and then 'Devices.'

Devices

Bluetooth, printers, mouse

On the 'Bluetooth and other devices screen' under the 'Other devices' heading, you should see a new COM port with a rather catchy name. Take a note of the COM port number.

 Silicon Labs CP210x USB to UART Bridge (COM8)

If the COM port is not there the PC is not seeing the radio. Check that the USB cable from the PC is plugged into The USB port on the back of the radio. Try unplugging the USB cable from the PC end and then plugging it back in again. If it still won't work, reload the Icom driver software with the radio plugged in but turned off.

➢ **The Audio Codec**

In addition to creating the two USB ports, the driver software creates an audio CODEC (coder-decoder). This makes the radio look to the computer like a sound card or an audio device like a microphone or speakers.

In Windows 10 select 'Settings' and then 'Devices.'

Audio

Devices
Bluetooth, printers, mouse

◁)) Speakers (Realtek High Definition Audio)

◁)) USB Audio CODEC

You should see the Icom audio codec listed as 'USB Audio CODEC.' If it is not there, the sound won't work. Try unplugging the USB cable from the PC end and then plugging it back in again. If it still won't work, reload the Icom driver software with the radio plugged in but turned off. It worked for me.

If you look at the sound card or mixer settings in your digital mode software, you should see a device labelled as a USB Audio CODEC or something similar. I found the labelling confusing, so I changed the names.

To change the names, in Windows 10, select 'Settings' and then 'Devices' as above. Then click 'Sound Settings' on the right side of the page. On the next window click 'Sound control panel,' on the right side of the page.

- On the 'Playback' tab you will see the Icom CODEC listed as a USB CODEC. Right-click the icon and select properties. You will be able to change the name and if you like to choose a different icon.

I picked the one that looks like a radio with a blue dial. The Playback tab is for sound out of the PC and into the radio. I changed the **Playback** device to '**Icom IC-7300 Input.**'

- On the 'Recording' tab you will see the Icom CODEC listed as a USB CODEC. Right-click the icon and select properties. You will be able to change the name and if you like to choose a different icon. I picked the one that looks like a radio with a blue dial. The Recording tab is for sound out of the radio and into the PC. I changed the **Recording** device to '**Icom IC-7300 Output.**'

➤ **Radio and COM Port device setting**

This is very important and can be a limitation on whether your digital mode or other PC software can "talk" to your IC-7300. Icom uses an address of 94h (Hexadecimal) for the IC-7300.

Some earlier Icom radios used 88h and most radios don't use an address at all. The IC-7610 uses an address of 98h. You can change the address in the radio using the CI-V settings, but this would be a last resort because it could cause software that is expecting to use 94h to fail.

What all this means is;

- If your digital mode software includes a device called Icom IC-7300. Select that option.

- A setting of IC-7600, IC-7610, IC-7800, or IC-9100 will not work. Probably because of the 94h address issue.

➤ **MixW**

Under <Hardware> <CAT Settings>, select CAT = 'ICOM' and Model = 'Other.' Set the 'Addr' to 94h then push OK. The address seems to change randomly after that, but it does not seem to matter. However, if you change any of the settings on the tab, make sure that the address is set back to 94h before you close the tab.

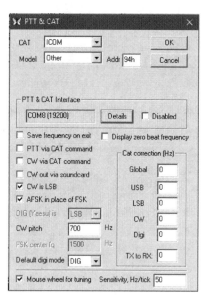

I set the Com port to COM8, 19200, 8, N, 1, RTS = PTT, DTR = CW.

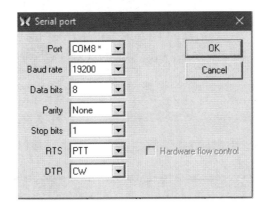

> ➤ **MMTTY**

For MMTTY click the 'Option(O)' tab. Select the 'Setup MMTTY(O). Then select the 'TX' tab and then the 'Radio Command' button on the right side. Set 'Port' to COM8, 'Baud' to 19200, Char wait to 0, Data length to 8 bits, 1 stop, No parity, and uncheck Xon/Xoff, check the DTR/RTS PTT box. Ignore the Commands box. But set xx to 94 and Model to Icom CI-V. Group is set to Icom xx=addr 01-7F.

Save the settings using the Save button, then exit using the OK button.

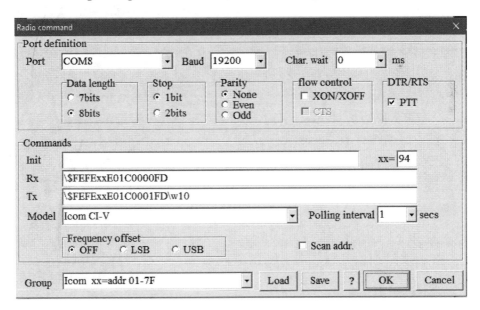

> ➢ **WSJT-X**

On WSJT-X v1.9.1 or later, select <Settings> <Radio> and choose the Icom IC-7300. Set the COM port (COM8) and the Baud Rate (19200). Leave Data Bits, Stop Bits, and Handshake set to 'Default.' Leave 'Force Control Lines' blank. This is important! Set 'PTT method' to RTS. Set 'Mode' to Data/Pkt. Set Split Operation to 'Rig.'

Press the test CAT button. It should go green to indicate that the software can communicate with the radio.

Exit the setup screen using the OK button.

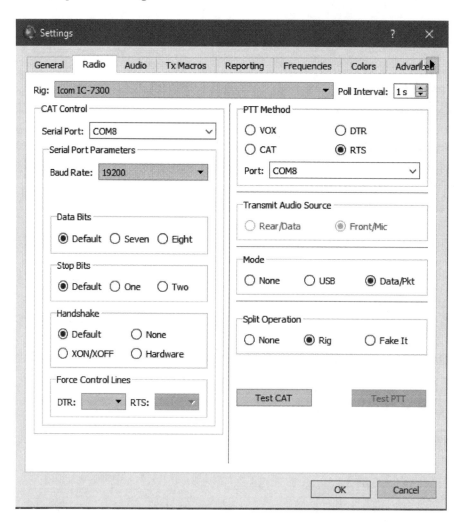

> **MRP40 CW program**

For MRP40, select Options, TX settings, Edit COM port pin configuration. COM8, Send pin = DTR, PTT pin = RTS, Activate PTT pin while sending via soundcard checked. Disable PTT Function is unchecked.

> **N1MM Logger+**

I am using N1MM Logger+ V 1.0.7300. It has support for the IC-7300. To set up communications with the radio, Start N1MM. Under 'Config - Configure Ports, Mode Control, Audio, Other - Hardware.' Set the Port dropdown list to the COM port (COM8). Set the Radio dropdown to IC-7300. Check CW/Other.

Click SET. I have speed: 19200, Parity: N, Bits: 8, Stop: 1, DTR: CW, RTS: PTT, Icom code 94h, Radio: 1, Delay: 30ms.

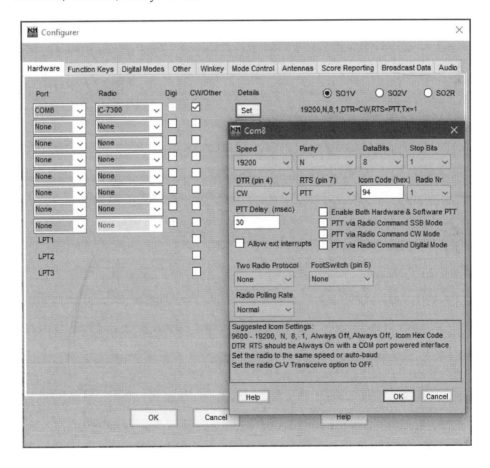

➢ **COM Port settings in the IC-7300 / CI-V settings**

The COM port settings in the radio are set in the CI-V menu. After some experimentation, I found that the default settings work well. I recommend leaving everything set to default. There is a full table of CI-V settings in the chapter on 'Special SET menu items' on page 118.

Note that WSJT-X will fail to communicate with the radio unless you set 'CI-V USB Echo Back' to 'ON' or set CI-V USB Port to 'Link to Remote.' Otherwise, you will get a *'Rig Failure'* error, labelled *'Protocol error while getting current VFO frequency.'* So, if you plan to use WSJT-X for FT8 or JT65 modes. Select <MENU> <SET> <Connectors> <CI-V> <CI-V USB Echo Back> and set it to ON.

➢ **COM Port settings in PC software**

The COM port settings for the PC are set in the digital mode or other PC software. They can also be set in Device Manager, but this does not seem to be necessary. Assuming you left the radio set for Auto, you can set the baud rate to anything up to 19200 bauds. These are the settings I use.

- Port   COM 8 (or the nominated CI-V com port)
- Baud Rate   19200
- Data bits   8
- Parity   None
- Stop bits   1
- RTS   PTT (the same as the USB Send/Keying setting)
- DTR   CW (the same as the USB Send/Keying setting)

➢ **USB SEND/Keying settings**

The radio creates a virtual COM port over the USB cable to the PC.

- The CI-V menu settings change the COM protocols, baud rate, start and stop bits etc. Everything can be left at default settings.

- The USB SEND/Keying menu sets the COM port control lines that PC software uses to change from receiving to transmitting mode and for CW keying. It also sets the RTTY keying line if you are using FSK (frequency shift keying using a digital keying signal) rather than AFSK (frequency shift keying using audio tones).

Select <MENU> <SET> <Connectors> <USB SEND/Keying>

- USB SEND is the PTT transmit line. I set it to 'RTS.'

- USB Keying (CW) is the line used to send the CW characters once the send line has switched the transceiver to transmit. I set it to 'DTR.'

- USB Keying (RTTY) is the line used to send the RTTY FSK characters once the send line has switched the transceiver to transmit. I set it to 'DTR.'

- Inhibit Timer at USB Connection. This is used to stop the radio sending a SEND, transmit keying signal when the USB cable is first plugged in.

  The manual states that this is only a problem if you are using old firmware. But I left it set to ON.

The RTS and DTR labels don't matter. You can use either line for the transmit PTT as long as you use the other line for CW. The names relate to old-fashioned RS-232 communications between 'old school' computers. RTS stands for 'ready to send' and DTR stands for 'data terminal ready.' But the lines have not been used for that sort of signalling since the 1970s. Anyway, I always use 'ready to send' for the 'Send' command and 'data terminal ready' for the CW data signal.

➢ **Windows PC soundcard settings**

The previous section deals with setting up the USB cable connection between the radio and the PC. The Icom USB driver software installs an Audio CODEC which acts like a soundcard in the computer, connected to the receiver audio output and transmitter audio input. Once the driver installation has been completed you should configure the Windows soundcard settings for the new audio codec.

Windows 10 treats the radio output as a mono 'microphone' and the radio input as stereo 'speakers' I found that confusing, so I renamed the 'recording device' (sound out of the radio and into the PC) to 'Icom IC-7300 Output' and I changed the 'playback device' (sound out of the PC and into the radio) to 'Icom IC-7300 Input.'

The radio codes and decodes the audio streams at a sampling rate of 48,000 Hz. It is best, but not essential, to set the Windows sound card to the same sampling rate.

In Windows 10 select 'Settings' and then 'Devices.' At the right of the page click 'Sound Card Panel'.

- On the 'Playback' tab, select the Icom USB CODEC. Windows will probably label it as a line card or speakers. I renamed it to 'Icom IC-7300 Input.' Right-click the mouse and select Properties. You can change the name and icon on the General Tab. Then select Advanced. Use the drop-down control to select '16 bit, 48000 Hz (DVD Quality)' and then click OK.

- On the 'Recording' tab, select the Icom USB CODEC. Windows will probably label it as a microphone. I renamed it to 'Icom IC-7300 Output' and changed the icon to one that looks like a radio. Right-click and select Properties. You can change the name and icon on the General Tab.

  Then select Advanced and use the drop-down control to select '2 channel, 16 bit, 48000 Hz (DVD Quality).'

- To Exit Click OK and OK again.

➢ **IC-7300 audio settings - ACC/USB Output Level**

I have chosen to leave the audio level being sent to the PC at the default 50% setting and will adjust the audio levels on the PC if required. If you do want to change the audio level being sent to the PC, select <MENU> <SET> <Connectors> <ACC/USB Output Level> on page 1/4. [*Make sure you don't choose the ACC/USB IF Output level on page 2/4 by mistake.*]

➢ **IC-7300 audio settings - USB MOD Level**

To avoid the possibility of overdriving the transceiver on digital modes, I have chosen to reduce the audio level being used to modulate the transmitter so that the RF output power is just peaking to 100 Watts. The setting should only have to be done once although you will have to go through the procedure several times to get it right. After the radio is set to your satisfaction you can adjust the PC soundcard output level or the transmit level control on the digital mode software. To make all of you digital mode software output the correct level. That way you can swap programs without having to worry about MOD (modulation) levels.

Unfortunately, you can't display the USB MOD Level menu setting and the transmitter metering at the same time, so you have to set the modulation level, then transmit a digital mode while observing the metering. Then go back to the menu to make an adjustment to the MOD level and then back to the meter display while you transmit again. Eventually, after a few cycles, you will end up with the level set perfectly.

- Before you begin, set the transmitter for 100% power using the MULTI control.
- To set the audio level being sent to the transmitter modulator, select <MENU> <SET> <Connectors> <MOD input> <USB MOD Level>. Set it for about 35%.
- Touch and hold the meter scale or press <MENU> <METER> to show the multi-function meter. While transmitting your favourite digital mode (I used FT8), check the RF Power output meter and the ALC meter.

- The level is set correctly when the transmit power is between the red +60 dB point and the end of the scale indicating full RF power output. The ALC should be showing one or two blue bars at the most. Any more than that and the ALC will be controlling the transmitter's output level which is not desired for digital modes. The low ALC setting ensures that nothing is being overdriven along the transmitter audio chain. The RF Power reading indicates that the digital mode is sending at full power. If you want to reduce power, you can turn down the RF Power using the MULTI control.

- I ended up with the USB MOD Level set to 34%. Your setting may be different depending on the settings in your PC and digital mode software. I have adjusted all of my digital mode programs, MRP40, MixW, and WSJT-X so that they all transmit at the same level. i.e. full RF power and two bars of blue on the ALC meter.

SETTING UP FOR FT8 OR OTHER DIGITAL MODE SOFTWARE

To use a PC software digital mode program for FT8 or any other external digital mode program, you will first have to establish communication between the PC and the radio. You can use the old CI-V interface via the REMOTE jack and send audio through an interface box to the radio, but that is terribly old-fashioned, not to mention relatively difficult. So, I will describe the easy way, using a USB cable. You will also have to set up an audio connection and set the audio levels. These steps are covered in the previous sections. Generally, this only needs to be done once and any differences between PC programs can be managed by changing settings on the PC.

➢ **First steps**

First set up a connection between the radio and the PC. Follow the steps in the 'Setting up a connection with your PC' section starting on page 16.

Then set the Windows PC sound settings on page 23 and the USB port audio settings on page 24.

That's it! You are finished with the setup. Unless you are planning to use WSJT-X for FT8. If that is the case, you need to check two more things.

➢ **IC-7300 CI-V setting for WSJT-X**

If you plan to use WSJT-X for FT8 there is one setting that **must** be made on the radio. WSJT-X will fail to communicate with the radio if you don't set CI-V USB Echo Back to ON. You get a *'Rig Failure - Protocol error while getting current VFO frequency'* error message.

Select <MENU> <SET> <Connectors> <CI-V> <CI-V USB Echo Back> and set it to ON. All of the other CI-V menu items can be left at default settings.

➤ **WSJT-X software COM settings**

There is an excellent WSJT-X user guide which tells you all about operating FT8 and the other WXJT modes. It is at http://physics.princeton.edu/pulsar/K1JT/wsjtx-doc/wsjtx-main-1.9.1.html#_standard_exchange.

Here are the settings that you will get you started with WSJT-X.

- On the WSJT-X software, go to 'File' 'Settings' 'General' and set your callsign, six-digit Maidenhead grid, and IARU region.

- Go to 'File' 'Settings' 'Radio' and set Rig to Icom IC-7300.

   o Set the serial **port** (usually COM8), **Baud rate** can be anything. I used 19200.

   o **Data bits**: Default or Eight, **Stop bits**: Default or One, **Handshake:** Default or None

   o Leave the Force Control Line blank

   o On the right side, set the 'PTT method' to RTS. Set the 'Mode' to Data/Pkt. Set 'Split' to None

   o Click 'Test CAT.' The button should turn green.

- Go to 'File' 'Settings' 'Audio'

   o Set the Audio input to your IC-7300 audio output. Leave the other setting at Mono.

   o Set the Audio Output to your IC-7300 audio input. Leave the other setting at Mono.

- Click OK to exit.

At this stage, the radio frequency should be indicated on the WSJT display and it should change if you turn the radio's main VFO knob. You should be able to click the band dropdown beside the frequency display and select a band. The radio should follow your selection.

➤ **RF Power in digital modes**

For the FT8 digital mode, which has relatively short transmitting periods, I have not experienced any problems running 100 Watts.

The temperature meter does not rise to any significant level. Touch and hold the meter scale or press <MENU> <METER> to show the multi-function meter. The temperature meter is at the bottom right.

However, for other digital modes, if you are prone to very long 'overs' I suggest monitoring the multi-function temperature meter. If the temperature meter does not get up to 'hot' you can run 100% transmit power. If the transmitter is getting hot, reduce the RF power to 75% by turning the Multi knob.

THE FT8 PRESET

The February 2021 (V1.40) firmware update added a "one-touch FT8 Preset" to the second page of the main menu. It adds five one-touch "mode pre-sets." The idea was to quickly change the radio to the settings required for FT8, reflecting the popularity of the mode. The top pre-set is labelled 'Normal,' but I have renamed it to 'SSB' because it does <u>not</u> return the radio to the previous non-Preset setting. The second item is labelled 'FT8,' although the options it sets are suitable for most external digital modes. You can edit those two settings and add three more pre-set arrangements of your own design.

Load the FT8 Preset before you start WSJT-X or another digital mode program. <MENU> <2> <PRESET> <FT8> <YES>

Unload it again when you have finished with the FT8 or other Preset. <MENU> <2> <PRESET> <UNLOAD> <YES>

Tip: one of the things that you can add to a 'Preset' is the Icom CI-V address. This is very handy if you have digital mode software that does not support the IC-7300 and needs a different CI-V address. Simply make a 'Preset' profile with the CI-V address that the digital mode software expects. For example, the IC-9700 uses A2h. When you 'Unload' (turn off) the 'Preset' the CI-V address will return to the normal 94h setting.

When you have finished using the Preset, touch the 'UNLOAD' Icon. Do not select 'Normal' because that is just another Preset, and it may well be different from your normal settings.

➢ **To add a Preset**

You can save the current radio settings as a new 'Preset,' <MENU> <2> <PRESET>. Touch and hold a vacant 'Preset' slot, or you can overwrite an existing 'Preset. Then select <Save to the Preset Memory>.

Add a Preset Name. Make any other changes you need, then scroll down to page six and select <<Write>> <YES>.

➢ **Editing the Preset settings**

You cannot edit a 'Preset' if it is in use. Touch UNLOAD first, <MENU> <2> <PRESET> <UNLOAD> <YES>.

Touch and hold the 'Preset' slot you wish to edit. Then select <Edit the Preset Memory>. Make any other changes you need, then scroll down to page six and select <<Write>> <YES>.

Each PRESET can store, a Preset name, the mode, receiver filter, filter bandwidth, USB keying settings, CI-V settings, data mod type, data off mod type, TX bandwidth, speech compressor, and TX wide/mid/narrow.

You can select items that you want to store and unselect any irrelevant items. For example, I set up a Preset for FM receiving. It does not need any of the CI-V or transmitter settings.

➢ **Turning off the Preset**

A Preset set on one band will not apply if you change to a different band. But if you return to the band with the Preset, it will still be in place.

If you change any of the settings, such as changing mode, the radio automatically unloads the Preset. If you change the mode back again, the Preset settings will return. The permanent way to turn off a Preset is to use <MENU> <2> <PRESET> <UNLOAD>.

➢ **Preset settings**

Adding a preset automatically saves the following settings. They cover pretty much everything you are likely to change when switching to digital mode operation.

Name of the preset	ACC/USB IF Output Level	USB Keying (RTTY)
Mode	USB MOD Level	USB Inhibit Timer
Filter	DATA MOD	USB Serial Function
Filter BW	SSB-D TX Bandwidth	CI-V Address
Filter type (HF)	DATA OFF MOD	CI-V Baud Rate
Filter type (50 MHz)	COMP	CI-V Transceive
ACC/USB Output Select	SSB TBW	CI-V USB Port
ACC/USB AF Output Level	USB SEND	CI-V USB Baud Rate
ACC/USB AF SQL	USB Keying (CW)	CI-V USB Echo

SETTING UP THE RADIO FOR RTTY OPERATION

The radio supports three kinds of RTTY operation. Firstly, there is the onboard RTTY decoder. Which can be used with the RTTY message memories. In this mode, you can take advantage of the excellent TPF (twin passband filter). I recommend using the TPF filter all the time because it really helps with accurate decodes. The second method is to use external PC software such as; MixW, MMTTY, MMVARI, Fldigi etc. with AFSK (audio frequency shift keying). AFSK uses two audio frequencies to create the frequency shift keying in the SSB mode. The third method is the FSK mode which uses a digital signal to key the transceiver to predefined mark and space offsets.

➤ **Onboard RTTY operation**

Select the RTTY mode. If the radio is already on RTTY it will select RTTY-R. This changes the operating sideband and more importantly it reverses the Mark and Space frequencies. If decode is gibberish the other station might be transmitting on the other sideband, creating RTTY-R.

You can change the RTTY MARK tone frequency and the RTTY shift. The default is that the Mark tone is at 2125 Hz and the Mark/Space shift is 170 Hz. In RTTY-R the Space tone is at 2125 Hz and the Mark/Space shift is still 170 Hz.

You can change the settings, but I don't see any reason to do so.

- <MENU> <SET> <Function> <RTTY Mark Frequency> 1275, 1615, or 2125
- <MENU> <SET> <Function> <RTTY Shift Width> choose 170, 200, or 425
- <MENU> <SET> <Function> <RTTY Keying Polarity> Normal or Reverse

➤ **Keyer send messages (keyer macros)**

There are eight RTTY message menus which can be used for DX or Contest operation or even just to save you sending the same message over and over. They are great for sending CQ on a quiet band. At a minimum, you will have to add your callsign, or the macros will send 'Icom' in place of it.

To edit the messages, put the radio into RTTY mode and select <MENU> <DECODE> <TX MEM> <EDIT>. Touch the message that you want to edit. Then select EDIT. An onscreen keyboard will appear, and you can edit or replace the message. Each message can be up to 70 characters long.

The ↵ symbol is a carriage return, it creates a new line on the text display at the receiving station.

Make sure that you press ENT, or your changes will be lost when you exit the keyboard screen.

Touch the Return icon ᔤ or the EXIT button to exit each menu layer.

To send the messages, while in the RTTY mode, select <MENU> <DECODE>. Then <TX MEM> and a message RT1 to RT8. After the message has been selected, the display reverts to the decode screen.

➢ **RF Power in RTTY mode**

You can run 100 Watts, but if you are prone to long 'overs' I suggest derating the power to 75 Watts. Press MULTI select the RF POWER icon and reduce RF power to 75% by turning the Multi knob. Keep an eye on the temperature meter on the multi-function meter display. If the transceiver is running hot, de-rate the transmitter power.

➢ **AFSK RTTY audio levels**

When using the built-in RTTY mode the modulation level is not user adjustable. If you are using an external digital mode program the levels should be set as per the Windows PC sound settings on page 23 and the USB port audio settings on page 24. You can set the audio levels in the IC-7300, or you can set the levels using the soundcard mixer controls in your PC or possibly with level controls in the digital mode software.

I have chosen to leave the audio level being sent to the PC at the default 50% setting and will adjust the audio levels on the PC if required. If you want to change the audio level being sent to the PC, Select <MENU> <SET> <Connectors> <ACC/USB AF Output Level>.

I have chosen to reduce the audio level being used to modulate the transmitter so that the RF output power is just reaching 100% and the ALC is not more than two blue bars. That ensures that nothing is being overloaded along the audio chain.

To change the audio level from the PC to the transmitter, select <MENU> <SET> <Connectors> <USB MOD Level>. I have my USB MOD Level set at 34%.

➢ **FSK RTTY from an external PC program**

I normally use either the internal RTTY system or the MixW PC software running RTTY in AFSK mode, rather than using FSK. However, one advantage of using FSK rather than AFSK from an external PC program is that you use the RTTY mode on the radio rather than the USB-D data mode. That means that you can use the TX MEM messages and the TPF (twin peak filter).

The setup for the radio is unchanged. First set up a connection between the radio and the PC. Follow the steps in the 'Setting up a connection with your PC' section starting on page 16. Then set the Windows PC sound settings on page 23 and the USB port audio settings on page 24.

If you want to run FSK RTTY and set up with the N1MM logger using MMTTY, refer to the notes by K0PIR at http://www.k0pir.us/icom-7300-rtty-fsk-mmtty/ or a video on the topic at https://www.youtube.com/watch?v=NmNHVjjAdiY.

You can set MMTTY for RTTY via FSK, but you can't use CI-V CAT control at the same time unless you run the CI-V control over the REMOTE jack using a second cable. Or you split the COM port using Eterlogic VSPE (Virtual Serial Port Emulator) software. See http://www.cedrickjohnson.com/2017/03/icom-ic-7300-usb-for-radio-control-fsk-keying/ for Cedrick Johnson's explanation.

I have not tried the split COM port method because you don't really need CI-V CAT control for MMTTY anyway. But you might need a second COM port or method of control if you want to run N1MM+ logger and MMTTY together.

> **MMTTY Setup for FSK RTTY**

1. On the MMTTY program click Option(O)

2. Select Setup MMTTY(O)

3. Click the TX tab and then the Radio command button.

4. Set the Port to NONE or to a port that is to be used for CI-V control. It cannot be set to the same port as the one you are using for FSK.

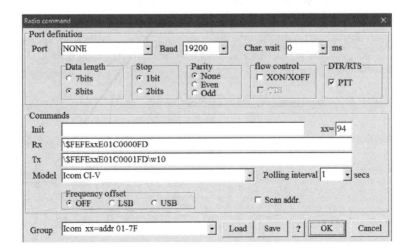

5. Back on the TX tab select EXTFSK64 in the PTT & FSK dropdown box. This
    is the Windows 10 setting. For 32 bit Windows or other 32 bit operating
    systems use the EXTFSK option.

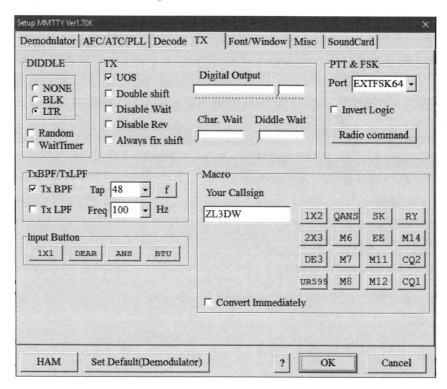

6. Selecting the EXTFSK64 (or EXTFSK) option will open a small popup
    window. Sometimes it ends up under other windows especially if you
    neglected to close it previously. Set the Com port for FSK operation. In my
    case, since I am using the standard USB connection to send FSK it will be
    COM8. Set output PTT to RTS and FSK output to DTR. This must match the
    CI-V settings in the radio.

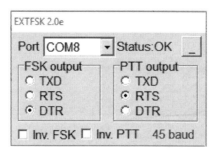

7. On the Misc tab change TX Port to COM-TxD(FSK). This makes the application use the DTR Com port control line to send a digital FSK signal rather than 'Sound' which sends AFSK tones.

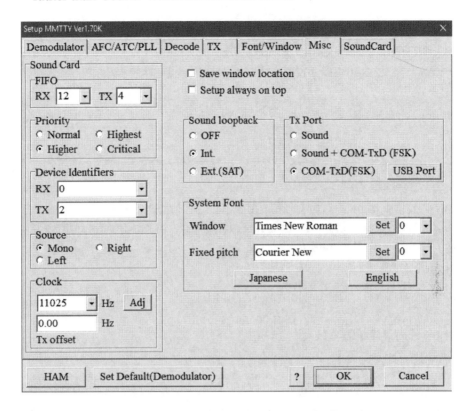

SETTING UP THE RADIO FOR PSK OPERATION

You can use an external PC software such as; MixW, MMTTY, MMVARI, Fldigi etc. to send and receive PSK. There is no internal PSK decoder.

To use external PC based digital mode programs you must select the USB-D DATA mode. First, ensure that radio is in SSB mode by touching the blue mode Soft Key and then touch DATA. The mode should change to USB-D. (Even below 10 MHz).

*Note that audio is sent to the PC when it is in any mode, so you can use your digital mode PC software to see and decode PSK signals. But the radio **must** be in the USB-D DATA mode to transmit PSK from your digital mode PC software.*

To use PC software to control the radio you have to connect a USB 2.0 Type B to Type A cable between the USB port on the rear of the radio and a USB 2.0 (black) or USB 3.0 (blue) port on the PC. If you have a spare USB 2.0 port on the PC, feel free to use it as there is no advantage in wasting a USB 3.0 port.

Using the USB cable for the first time is **not** "plug-n-play." If you haven't already loaded the Icom driver and set up the USB ports and audio codec yet, go to 'Setting up a connection to your PC' on page 16.

➢ **PSK audio levels**

Once the digital mode software can communicate with the radio, it is time to set the audio levels on transmit and receive. If the audio levels have already been set for RTTY or other digital modes, you are done. Go and work some PSK stations.

If you have already set up the DATA mode audio levels, you can skip this section.

You can set the audio levels in the IC-7300, or you can set the levels using the soundcard mixer controls in your PC or possibly with level controls inside the digital mode software.

For the IC-7300, I have chosen to leave the audio level being sent to the PC at the default 50% setting and will adjust the audio levels on the PC if required. If you do want to change the audio level being sent to the PC, Select <MENU> <SET> <Connectors> <ACC/USB AF Output Level>.

Also set <MENU> <SET> <Connectors> <ACC/USB AF Output Select> to 'AF.'

I have chosen to reduce the audio level being used to modulate the transmitter so that the RF output power is just reaching 100% and the ALC is not more than two blue bars. That ensures that nothing is being overloaded along the audio chain. However, PSK is a full power mode, so if you are prone to long 'overs' I suggest derating the power to 75 Watts using the RF Power control on the MULTI menu. Keep an eye on the temperature meter on the multi-function meter display. If the transceiver is running hot, de-rate the transmitter power.

 To change the audio level from the PC to the transmitter, select <MENU> <SET> <Connectors> <USB MOD Level>. I have mine set at 34%, but it depends on the Windows sound settings and the digital mode software audio settings.

The settings are more fully covered in the 'Display Soft Keys' chapter in the 'Audio levels (USB cable) section on page 120. But all you really need are those two adjustments.

SETTING UP FOR FM REPEATER OPERATION

Repeater operation is achieved using the Split mode. Select FM mode and set the Main VFO to the repeater output frequency.

Press and hold the SPLIT button to enable 'Quick Split.' The split mode will be turned on and so will CTCSS tone encoding, (whether you want it or not).

The Split indicator is under the time display and the Tone indicator is under the filter icon between the mode and the FIL setting. The sub VFO (VFO-B) display will show the transmit frequency for the repeater input. Holding the XFC 'transmit frequency check' button lets you temporarily listen on the repeater input frequency and it also replaces the sub VFO frequency with an indication of the split in kHz.

➢ **Setting the FM split**

You can set the pre-determined split offset for the 6m band (50 MHz) and for the rest of the HF bands.

- **HF bands**: \<MENU\> \<SET\> \<Function\> \<SPLIT\> \<FM SPLIT Offset (HF)\> (Default -100 kHz)

- **6m band:** \<MENU\> \<SET\> \<Function\> \<SPLIT\> \<FM SPLIT Offset (50M)\> (Default -500 kHz)

- New Zealand 6m repeaters use a -1 MHz offset.

➢ **Setting the TONE**

The TONE function Soft Key \<FUNCTION\> \<TONE\> is only visible when the transceiver is in FM mode. TONE is used for repeater operation primarily on the 6m band. The Soft Key has three modes: OFF, TONE, and TSQL. A blue indicator indicates that tone is in use.

- The TONE mode sends a tone with your FM transmission that opens the repeater squelch.

- The TSQL mode squelches your receiver until the correct tone is received from the repeater. It also sends the tone on your transmission to open the repeater squelch.

- The OFF mode turns off both receiver tone squelch and the transmitted tone.

Touch and hold the TONE Soft Key to open the setup screen.

- REPEATER TONE is the tone that is sent to the repeater in the TONE mode.

- T-SQL is the tone that must be received from the repeater to open the receiver squelch in the TSQL mode. The same tone is sent with your transmit signal.

- Select the option you want to change by touching the appropriate Soft Key and rotate the VFO knob to set the tone frequency that you want.

The full set of CTCSS tones are available. Most repeaters use 67 Hz or the Icom default tone of 88.5 Hz.

Touch and hold DEF to reset the tone back to the default 88.5 Hz.

CTCSS Tones (Hz)				
67.0	69.3	71.9	74.4	77.0
79.7	82.5	85.4	88.5	91.5
94.8	97.4	100	103.5	107.2
110.9	114.8	118.8	123.0	127.3
131.8	136.5	141.3	146.2	151.4
156.7	159.8	162.2	165.5	167.9
171.3	173.8	177.3	179.9	183.5
186.2	189.9	192.8	196.5	199.5

There is a useful icon called T-SCAN on the TONE FREQUENCY sub-menu.

- To find the tone that you should be transmitting for the TONE mode. Tune your receiver to the repeater input frequency then touch the REPEATER TONE icon. When another station is using the repeater, touch T-SCAN. The radio will scan through all the possible CTCSS codes until it finds the one that matches the tone on the other station's repeater input transmission. This method will probably work if the receiver is tuned to the repeater output frequency, as the same tone is probably transmitted from the repeater.

- To find the tone that you need for tone squelch in the TSQL mode. Tune your receiver to the repeater output frequency then touch the T-SQL - TONE icon. When another station is using the repeater, touch T-SCAN. The radio will scan through all the CTCSS codes until it finds the one that matches the tone on the repeater's output transmission.

➤ **Memory channels and FM**

There is no specific memory storage for FM channels. You can use the standard memory channels and name the memory channel with the repeater identification. The standard memory channels do store the Split offset so you have to remember to set that before saving the frequencies to a memory slot. You can also store a repeater output frequency in the MPAD memory pad store. But it does not store the Split offset.

Recalling a stored frequency is the same as for any other channel. In VFO mode use the up and down buttons to select the channel number and then press V/M to change to the memory channel mode.

To store repeater frequencies, you have to set the VFO to the repeater output frequency, then press and hold the SPLIT button. This will change VFO B to the default split for FM, (-100 kHz for the HF bands and -500 kHz for the 6m band). These offsets can be changed in the <MENU> <FUNCTION> <SPLIT> submenu.

When the frequencies, split mode and tone have been selected you can open the memory manager.

Method 1:

Use the up and down buttons to select a blank or any memory channel. Touch the memory number on the touch screen display to open the VFO/MEMORY submenu. This only works on the display that has a larger frequency display with the memory number at the right of the screen. Touch and hold MW to save the information to the memory slot.

Method 2:

Use <MENU> <MEMORY> to open the menu manager. Use the onscreen up down buttons, or the MULTI knob, or the front panel up down buttons, to select the memory slot you want to use. Touch and hold the position of the memory slot. You will be asked if you want to save the information to the memory slot. Touching the star underneath the channel number lets you add the slot to one of the three scan groups. Touching the 'file' icon at the light of the line, lets you edit (or add) a name for the channel.

CONNECTING AND USING A LINEAR AMPLIFIER

➢ **Linear amplifier connections**

If you have an Icom linear amplifier it will be controlled via an Icom cable connected to the 4 pin Molex connector on the rear panel.

If you have a non-Icom linear amplifier you would normally connect a cable from the SEND connector on the radio to the amplifier's PTT input and another from the amplifier's ALC output to the radio's ALC input. You can buy a standard stereo audio cable with RCA connectors. I always use the red connector for PTT and the white for ALC.

In my setup, the IC-7300 SEND line connects to the antenna tuner PTT input and there is a short cable from the tuner PTT output to the PTT input on the linear amplifier.

An ALC (automatic level control) connection between the linear amplifier and the transceiver is not essential but it is recommended. The cable goes directly from the ALC output on the amplifier to the ALC input connector on the IC-7300.

The ALC output from the linear amplifier turns down the transceiver output power in the event that the amplifier is being overdriven. It should be configured so that it is not operating unless the transceiver is accidentally left at full power when driving the amplifier. Don't use it as a method of controlling the amplifier power. It should only be used as a failsafe in the event of a power setting mistake. ALC also protects the linear amplifier from 'overshoot.' Some transceivers emit a full power RF spike when you key the transmitter even when the RF power is set to a low level. Without ALC control this has a tendency to trip the protection circuit in the amplifier.

➢ Band switching

My Elecraft KPA-500 amplifier has both manual and RF sensed automatic band switching. With a suitable cable, it can take advantage of the Icom band output signal to switch bands. The band output voltage is available on the rear panel ACC connector.

If your amplifier can't handle the Icom band output and you really want to use automatic band switching, Ronald Rossi, KK1L describes a kit for converting the Icom band output to discrete band switch, outputs.
See http://kk1l.com/kk1l_2x6switch/KK1LIcomDecoderInstructions2d2.pdf.

Another option is the band decoder from RemoteQTH.com
https://remoteqth.com/arduino-band-decoder.php.

Or the Bob Wolbert K6XX design at http://www.k6xx.com/radio/icbsciv.pdf.

➢ Setting the ALC level

Find out how to adjust the ALC output level on your amplifier. It might be a menu setting as in the case of my Elecraft amplifier or it might be a control on the front or the rear panel of the amplifier. Press MULTI on the radio and turn the RF Power level down to about 20%. If your linear amplifier needs a lot of drive, the required output may be higher. Talk into the microphone in your 'normal radio voice' or use CW or your digital mode of choice.

Remember to identify your station as you transmit. Increase the RF LEVEL control until the Linear Amplifier peaks to full power, (or the highest power that you want to run). Note the setting of the RF LEVEL control for future reference.

Now, while transmitting, increase the ALC level from the amplifier until the output power from the transceiver starts to decrease. Then back the control off a little so that full power is being generated and the ALC is not having any effect on the output power. That should be the correct setting.

If you increase the RF LEVEL from the Icom transceiver the ALC voltage should hold the output power down so that the Linear Amplifier will not be driven into an overload situation.

The ALC control is an amplifier protection method, like a circuit breaker or a fuse. You can get spurious outputs and intermodulation if you intentionally run full power from the radio and rely on the ALC to limit the RF power. Always reduce the RF LEVEL control to the safe level that you determined above. That way the ALC should never operate, but it is there in case of "finger trouble."

➤ **Digital modes**

I can't find anything in the Icom documentation that suggests that you have to reduce the transmitted RF power for continuous transmission modes like FM, AM or digital modes. However, it is very likely that your amplifier is not designed for long periods running at full power. Check the amplifier manual and if necessary, down-rate the output for digital modes. I don't usually run more than half power on digital modes. It is not necessary for most contacts. High power on FT8 does not help all that much and it annoys other stations. Please try 100 Watts first and only use high power when you really need it.

USING AN SD CARD

You must have an SD card in the radio if you want to record signals off the air, store the radio configuration, store memory channel contents, RTTY decode logs, voice message memories, or saved screen capture images, or to upload new firmware. You do not need an SD card for the CW or RTTY message keyers.

The blue SD indicator in the top right of the display just to the left of the clock indicates that there is an SD card in the SD card slot. A flashing SD icon indicates that the radio is writing information to, or reading information from, the SD card.

You can use a 2 Gb SD card or an SDHC card from 4 Gb up to 32 Gb. Icom recommends using SanDisk© SD cards. I am using a 16 Gb SanDisk SDHC card.

The information screen says that after recording all of the keyer macros and saving the settings a few times that I have 275 hours of recording time left. So, buying a 16 Gb card, as I did, is probably overkill. Unless I am transferring images to my PC, I leave the SD card in the radio all the time.

Icom recommends formatting the SD card using the function in the radio, before using it for the first time. <MENU> <SET> <SD Card> <Format>. There is also a menu option to 'unmount' the SD card before you remove it. (The same as you would when removing a USB stick from your PC). <MENU> <SD Card> <Unmount>. I don't bother using it and haven't had any problems, but it's your risk.

FIRMWARE UPDATES

Updating the firmware is a bit scary because during the process you are presented with some bright yellow warning screens. However, I found that updating the firmware went very smoothly.

You need an SD card and a PC that can write to it. If your PC can't write to an SD card you will have to buy a USB SD card reader. Mine cost about $10.

➢ **Checking the installed firmware revision**

The currently installed firmware revision is displayed at the bottom right of the splash screen during the radio startup process. Or you can find it at <MENU> <SET> <Others> <Information> <Version>.

The display indicates the five firmware versions. The overall 'firmware version' is the Main CPU revision number.

- Main CPU: (version 1.40 or newer)
- Front CPU: (version 1.01 or newer)
- DSP Program: (version 1.07 or newer)
- DSP Data: (version 1.00 or newer)
- FPGA: (version 1.13 or newer)

➢ **Downloading firmware from the Icom website**

Visit the Icom website at http://www.icom.co.jp/world/index.html Click 'Support' then 'Firmware Updates/Software Downloads.' Find the IC-7300 section and look for the newest firmware release. Currently, it is '*Firmware (Version 1.40) and manuals.*' If the latest release is newer than the one currently installed on the radio, you should download the new firmware. A single file is used to update all four firmware updatable devices on the radio.

Double check that you are downloading software for the IC-7300. Click on the link and then 'Agree.' Click 'Save As' and nominate the location where you want to download the zip file. After the download has finished, find the downloaded file in the download directory. Right-click the file name and select 'Unzip.' Either unzip the file directly to the IC-7300 directory on the SD or unzip it to your hard drive and then copy the file across to the IC-7300 directory on the SD.

The unzipped file will be called something like '7300_140.dat.' It **must** be copied into the IC-7300 directory on the SD card.

Do not turn the radio off during a firmware update.

➢ **Firmware update process**

- Unmount the SD card from the PC and plug it back into the radio. Wait a few seconds until the drive is 'mounted.' Blue SD icon not flashing.

- Select <MENU> <SET> <SD Card> <Save Setting> and save the current radio configuration. You can select a file name or just go with the default 'time and date stamp' file name. Touch the down arrow to get to page 2/2.

- Touch <Firmware Update> then after reading both pages of the yellow screen, be brave and select <YES>

- A list with at least one firmware file will be displayed. Select the file with the highest revision number. If a file is not displayed, it means that the firmware '.dat' file is missing or not in the IC-7300 directory.

- After you have read the scary precautions on the yellow screen, be brave again, and touch and hold <YES> for one second. You have already saved the radio configuration file in step two... right?

- Follow the instructions on the transceiver screen. **Do not turn the radio off** or touch any controls.

- The IC-7300 will read and check the firmware file from the SD card and then write the Main CPU and DSP/FPGA firmware updates to the radio. Progress is displayed on the screen.

- When the update is completed, "Firmware updating has completed." is displayed and the IC-7300 will automatically restart.

- As the radio boots, you will see the new firmware revision number displayed at the bottom right of the splash screen during the radio startup process.

- After the radio starts it should be back to normal operation. You can check out the new firmware revision numbers at <MENU> <SET> <Others> <Information> <Version>.

- Your settings will probably be intact but if they are not, you can recover them using <MENU> <SET> <USB Flash Drive> <Load Setting> and selecting the file name you saved earlier. Temporary settings such as the MPad memories and the band stacking registers may be lost.

SETTING THE CLOCK

The radio has a clock display, the type that tells the time, not the CPU or ADC clocks. Generally, you set the clock to display the local time, but you could set it to display UTC (universal coordinated time). The clock is displayed at the top right of the screen. Touching the clock icon displays Local and UTC time, day, and date.

The date is not displayed on the screen, but it is used for LOG timestamping.

➢ **Setting the clock to local time.**

1. Select <MENU> <SET> <Time Set>

2. Select <UTC Offset> use the plus or minus keys to set the UTC Offset for your location. Remember to add daylight saving time if daylight saving is in use. Touch the Return icon ↺ to exit.

3. Select <Date/Time>

4. Select <Date> use the up-down keys to set the year, month, and day (in that order). Make sure that you touch the SET icon when you have finished, or your changes will not be saved.

5. Select <Time> and use the up-down keys to set the current local time in 24-hour format. Make sure that you touch the SET icon when you have finished, or your changes will not be saved.

6. Press the EXIT button or touch the Return icon several times ↺ to exit.

➢ **Setting the clock to UTC.**

NOTE that setting the clock to UTC will cause the RTTY log to use UTC rather than local time.

1. Select <MENU> <SET> <Time Set>

2. Select <UTC Offset> use the up-down keys to set the UTC Offset to zero (0).

3. Select <Date/Time>

4. Select <Date> use the up-down keys to set the UTC year, month, and day (in that order). The date may be different from the date at your place. Make sure that you touch the SET icon when you have finished, or your changes will not be saved.

5. Select <Time> and use the up-down keys to set the current UTC time in 24-hour format.

6. Press the EXIT button or touch the Return icon several times ↺ to exit.

SETTING AMATEUR BAND LIMITS

The radio won't allow you to transmit outside the amateur radio bands, but these vary between IARU regions and from country to country. The radio beeps if you tune the VFO across an amateur radio band edge. To some extent, you can set these band edges to suit the legal requirements for your country. Note that these band edges are not the same as the ones that you set for the Panadapter 'FIX' mode.

My radio has pre-set band limits for eleven amateur radio bands: 160m, 80m, 60m, 40m, 30m, 20m, 17m, 15m, 12m, 10m, and 6m. I was able to reduce the size of the 160m and 80m band to comply with the New Zealand band plan, but it seems that you can't add the LF bands or the MARS frequencies without a hardware modification to the radio. Basically, you can split bands up, but you can't add new ones.

The 'User Band Edge' settings are hidden. To change the band limits from the factory defaults you first have to change the 'Band Edge Beep' setting. In the default ON setting, the 'Band Edge Beep' makes the radio beep when you tune to a frequency above or below an amateur radio band. You can turn the beep off if you want to by using <MENU> <SET> <Function> <Band Edge Beep>.

To enable the User band Edge menu option, you must set <MENU> <SET> <Function> <Band Edge Beep> to <ON (User)> or <ON (User) & TX Limit>.

- The ON (User) option lets you change the frequency where the Band Edge Beep occurs. You must leave the ON (User) option selected if you want the out of band beep to sound at the frequencies you have selected in the table.

- The ON (User) & TX Limit option lets you change the frequency where the Band Edge Beep occurs and the transmit band limits. You must leave the ON (User) & TX Limit option selected if you want the frequencies in the table to apply. Otherwise, the transmit frequencies are set by the diodes on the 'Main Unit' board.

Once you have changed Band Edge Beep setting, you can select <MENU> <SET> <Function> <User Band Edge> and then the band that you want to change. You will see a list of the currently programmed amateur bands and the frequency limits. You can program up to thirty band edges.

There are some rules.

1. Touch and hold a band edge to Insert, Delete, or Set it to the Default setting.

2. You cannot have any band overlap any other band. If you want to split a band you have to edit or delete the old band first.

3.   You cannot enter a frequency that is outside of the pre-defined amateur bands which are set in hardware with diodes.

4.   The bands must be in order. Lowest frequency band first. So, if you insert a new band you have to do it in the right place. Insert adds a new slot above the one you touch and held. But you can only enter frequencies that are available. In other words, you have to have trimmed down another band to make room for your new one. As an experiment, I shortened the 20m band to 14.1 to 14.350 MHz and added a new band from 14.0 to 14.099 MHz. This exercise was pretty pointless as the radio lacks 'band up,' 'band down' buttons. But it proved that you can make a new band.

5.   Are you still sure you want to do this?

You can check the Icom Basic manual page 3-6 to 3-8 for a more detailed explanation of the data entry requirements.

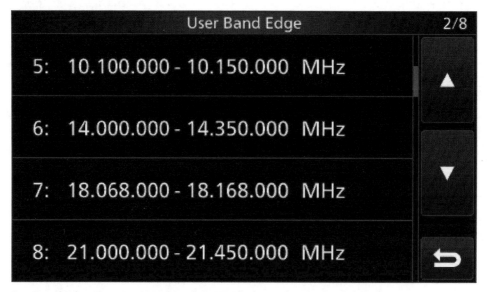

Set 'Band Edge Beep' to 'ON (User) & TX Limit' to change the band plan

Front panel controls and connectors

This chapter describes all the front panel controls. Generally, it provides some additional detail to supplement the information provided in the Icom manuals. Any menu options relating to the front panel controls are included.

POWER

I guess you have already worked this one out. The bottom right of the display shows the firmware revision during the boot-up process. And if you have entered your callsign it will display under Icom IC-7300 on the splash screen. Next, you get a display stating, 'RF Power (HF)' and your current RF Power setting.

Pressing the POWER button for a short time will store a screen capture of the current display image on to the SD card if the relevant menu settings have been enabled. <MENU> <SET> <Function> <Screen Capture [POWER] Switch> <ON>.

You can select to save in .png or .bmp graphics format. Both of these are uncompressed graphics formats. I chose the default .png setting. <MENU> <SET> <Function> <Screen Capture File Type> <PNG>.

Press and hold the POWER button for two seconds to turn the radio off.

TRANSMIT

The TRANSMIT button places the transceiver into transmit mode. Be careful to avoid transmitting full power for long periods of time. On the three voice modes, the microphone is live while transmitting. On FM the transceiver will transmit 100 Watts. On AM it will transmit 25 Watts. On SSB it will transmit normally for SSB. i.e. the power output is dependent on the microphone audio level. The TX indication on the main screen turns to white on red to indicate that the radio is transmitting.

TUNER

Pressing TUNER activates the automatic antenna tuner. A TUNE indication on the display just above the red TX indicator indicates that the tuner is active. The tuner will initiate a tune operation if the SWR exceeds 1.3:1.

Holding the TUNER button down forces a 'manual' tune operation. A carrier of about 10 Watts is generated and the TUNE icon turns red and flashes while the tuner is operating. You will hear relays clicking as the tuner attempts to find a match. If you just hear a single click and the TUNE icon turns red then white, the antenna is matched and the settings for that frequency have been memorised.

The tuner remembers the settings for each frequency that you transmit on, (to a 100 kHz resolution). So, after a while, you should not hear the relays clicking each time you transmit on a different area of the band.

If you press and hold the TUNER button a second time a full tune will be initiated.

There is a menu setting which makes the tuner active every time the PTT switch is operated. Set <MENU> <SET> <Function> <TUNER> <PTT Start> to ON.

VOX / BREAK-IN (VOICE MODES)

Pressing the VOX/BK-IN button while in a voice mode SSB, AM, FM, RTTY or one of the data sub-modes, will enable VOX (voice operated switch) transmit switching. VOX keys the transmitter when you talk into the microphone without the need for you to press the PTT switch on the microphone. It is handy if you are using a headset or a desk microphone without a PTT switch.

Press and hold the VOX/BK-IN button to enable the VOX MULTI menu. There are four settings. Touch the required setting and turn the MULTI knob to change the value. You should take the time to set the VOX up carefully as some settings tend to counteract other settings.

Press the press EXIT button or touch the Return icon ↺ to exit.

➢ **VOX GAIN**

VOX GAIN sets the sensitivity of the VOX. In other words how loud you have to talk to operate it and put the radio into transmit mode. (Default 50%).

➢ **ANTI-VOX**

ANTI-VOX stops the VOX triggering on miscellaneous noise like audio from the speaker or background noise. Higher values make the VOX less likely to trigger. (Default 50%).

➢ **DELAY**

DELAY sets the pause time before the radio reverts to receive mode. It needs to be set so that the radio keeps transmitting while you are talking normally but returns to receive in a reasonable time after you have finished talking. (Default 0.2 seconds).

➢ **VOICE DELAY**

VOICE DELAY, (Default OFF), sets the delay after you start talking before the transmitter starts. Generally, you want the radio to transmit immediately to avoid the first syllable or word being missed from the transmission. However, if you are prone to making noises perhaps you should set it longer. If you start each transmission with "Ah" you could set it quite long.

VOX / BREAK-IN (CW MODE).

Pressing the VOX/BK-IN button while in CW mode cycles through semi break-in (BKIN), full break-in (F-BKIN), or no break-in modes.

See 'break-in setting' on page 14.

Press and hold the VOX/BK-IN button while in CW mode to enable the MULTI Break-in display. There is only one adjustment. Turn the Multi knob to change the semi break-in delay setting. The default is a period of 7.5 dits at the selected CW keying speed.

PHONE JACK

The headphone jack is a 1/8" (3.5mm) stereo mini-phono jack which provides a mono output to both sides of a stereo headset or headphones. The impedance is a standard 8 - 16Ω. You could connect this jack to a set of amplified PC speakers.

The audio output to the headphone jack is affected by the AF Gain (volume) control and the squelch controls.

MIC JACK

The Mic jack is for the microphone. There is no facility for the connection of a balanced microphone. The jack is a standard Icom 8 pin connector.

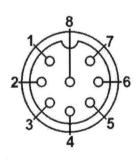

Pin	Description
1.	Microphone input (600Ω)
2.	+8 V DC output (max 10 mA) for electret microphones
3.	Up / Down buttons. 'Up' if grounded, 'Down' if grounded via a 470 Ω resistor. This pin can also be used to send message memories M1 to M4, see below.
4.	Squelch (goes low when squelch is open)
5.	PTT (pull low to activate transmitter)
6.	PTT ground
7.	Microphone ground
8.	Audio output (level controlled by AF control)

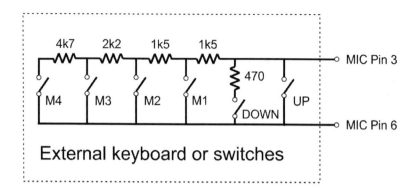

You can use the microphone jack to connect switches or a keypad used to send the first four of the eight keyer macros for the selected mode, (CW, voice, or RTTY). This might be useful if you have additional buttons on your microphone, or you wish to re-purpose the 'up' and 'down' buttons on the Icom Hand Mic.

To enable the external keypad or switches to trigger message transmission in the Voice, CW, or RTTY mode, you must set them to ON in the menu structure. This reduces the possibility of keying the transceiver by accident.

<MENU> <SET> <Connectors> <External Keypad> <VOICE>

<MENU> <SET> <Connectors> <External Keypad > <KEYER>

<MENU> <SET> <Connectors> <External Keypad > <RTTY>

➤ **Customised microphone buttons**

The UP and DOWN microphone buttons normally move the VFO frequency up or down by 50 Hz. If the Tuning Step function is on, the VFO moves by whatever the step size is, and in memory mode the buttons step through the memory channels.

Firmware release 1.40 (The February 2021) introduced the ability to change the function of the microphone buttons. You can allocate any of the 19 options to either of the two buttons. <SET> <Function> <MIC Key Customize>. The programmed buttons perform the same as the front panel buttons.

Function	Description
UP	(Default) 50 Hz step, or TS step, or memory channel
DOWN	(Default) 50 Hz step, or TS step, or memory channel
UP (VFO kHz)	1 kHz step, or TS step, or memory channel
DOWN (VFO kHz)	1 kHz step, or TS step, or memory channel

Function	Description
XFC	Hold to monitor transmit frequency (same as XFC button)
VFO/Memo	Push to swap VFO to memory channel mode. Hold to copy a memory channel into the VFO
BAND UP	Push to move up a band. Hold for the band stack
BAND DOWN	Push to move up a band. Hold for the band stack
SPEECH	Announce the frequency and mode, (S-meter optional)
MODE	Push to select a mode, hold to toggle USB to LSB, CW to CW-R or RTTY to RTTY-R
Voice/keyer/RTTY 1-4	Press to Send voice/CW/RTTY message 1-4 (dependent on the mode)
TS	Turn the tuning step mode on or off. Hold for TS screen
MPAD	Press to recall MPAD channel. Hold to save the current VFO settings to the MPAD
SPLIT	Press to toggle Split on or off. Hold for Quick Split.
A/B	Press to swap VFOs. Hold to save the hidden VFO settings to the currently displayed VFO
TUNER	Push to turn the tuner on or off. Hold starts a manual tune

TWIN PBT

Twin passband tuning is a standard feature of Icom receivers. It allows the operator to shift or narrow the IF passband to reduce the effect of very close interference signals. Note that this is a DSP function, not an analogue one. The topic is covered very well in the Icom Basic Manual page 4-3 so I will summarise.

The received RF signal passes through two filters which normally have identical passband responses.

Adjusting the black PBT knob shifts the passband of the PBT1 filter higher or lower in frequency. Adjusting the silver outer control shifts the passband of the PBT2 filter higher or lower in frequency. The effect that this is having on the IF passband is indicated in a pop-up diagram on the display. The popup disappears very quickly after stop adjusting the PBT controls. But there is a way to make visible for longer.

After adjustment, the resulting filter bandwidth is indicated on the small filter shape icon between the Mode and FIL icons at the top of the display. A tiny white dot to the right of the filter shape icon indicates that PBT is in use.

Push the black PBT knob to CLR (clear) the PBT filters back to normal.

It is much easier to understand the operation of these controls if you touch and hold the FIL icon at the top of the display screen. This will open the filter setup screen and allows you to observe the nifty colour display of the filter as you adjust the two PBT knobs. Press the press EXIT button to exit this screen.

P_AMP / ATT (PREAMPLIFIER OR ATTENUATOR)

The P.AMP/ATT button controls the status of the internal preamplifier and front-end attenuator. The button has the same function as the front panel P.AMP/ATT Soft Key on the FUNCTION menu.

Pressing the button cycles through; Preamp Off, Preamp 1, and Preamp 2. On the large VFO display, the selected choice is displayed underneath the VFO frequency. On the panadapter display, a popup shows the choice as you press the button.

Preamp 1 ($\approx$7 dB) is recommended for the low HF bands. Preamp 2 ($\approx$11-12 dB) is a higher gain amplifier recommended for the high HF bands. If signals are overloading the receiver, turn the preamplifier off.

Touch and hold turns enables a 20 dB attenuator.

NOTCH (NOTCH FILTER)

Pressing the Notch button cycles between the automatic notch filter, the manual notch filter, and off. An indicator is briefly displayed in the centre of the screen. On some screens, that is the only indication that a notch filter has been turned on. On the screen with the large VFO numbers, AN under the 10 Hz digit of the VFO indicates that the automatic notch filter is turned on. MN indicates that the manual notch filter is turned on.

Auto notch (AN) eliminates the effect of long-term interference signals such as carrier signals that are close to the wanted receiving frequency.

The manual notch (MN) can be placed anywhere across the receiver's audio passband, to reduce eliminate interfering signals. Holding down the Notch button opens the MULTI menu where you can select from wide, medium or narrow manual notch filters and set the notch frequency. Keep the filter as narrow as possible while effectively removing the interference.

➢ **The audio scope**

You can see the effect of the notch filter by opening the Audio Scope. Select <MENU> <AUDIO>. Tune to a frequency where you can hear a carrier 'birdie.' You must have the receiver squelch open to see signals on the audio spectrum scope. Turn on the auto notch and you will see the carrier signal disappear from the audio spectrum display and you won't be able to hear it anymore. Enable the manual notch and you will be able to see a black zone on the audio spectrum indicating a deep null in the signal. Changing from narrow to mid or wide makes the null zone wider. Adjusting the manual notch position moves the nulled band across the audio spectrum.

➢ **Menu settings**

There are two menu settings that can affect Notch operation. For AM and SSB you can select from Auto Notch only, Manual Notch only, or the default choice of both Auto and Manual notch options. Since the auto notch is best for interfering carriers and the manual option allows you to place the notch where you want it on the audio passband, I can't imagine why you wouldn't want both options. But you can change it if you want to. <MENU> <SET> <Function> <[NOTCH] Switch (SSB)> or <[NOTCH] Switch (AM)>.

NB (NOISE BLANKER)

Pressing the NB button turns the noise blanker on. On the display with the large VFO characters, Noise Blanker operation is indicated with a white NB indication just below the Hz digit of the VFO frequency. Other screens get a brief popup indication. Holding the NB button down brings up a sub-menu where you can adjust the Level, Depth, and Width. The noise blanker is designed to reduce or eliminate regular pulse-type noise such as car ignition noise. You may need to experiment with the controls when tackling a particular noise problem.

The LEVEL control (default 50%) sets the audio level that the filter uses as a threshold. Most DSP noise blankers work by eliminating or modifying noise peaks that are above the average received signal level. They usually have no effect on noise pulses that are below the average speech level. Setting the NB level to an aggressive level may affect audio quality.

The DEPTH control (default 8) sets how much the noise pulse will be attenuated. Too high a setting could cause the speech to be attenuated when a noise spike is attacked by the blanker. This could cause a choppy sound to the audio.

The WIDTH control (default 50%) sets how long after the start of the pulse the output signal will remain attenuated.

Set it to the minimum setting that adequately removes the interference. Very sharp short duration spikes will need less time than longer noise spikes such as lightning crashes.

The noise blanker is disabled when the radio is in FM mode.

Noise blanking occurs very early in the receiver DSP process. It is performed on the wideband spectrum before any demodulation or other filtering takes place. Noise reduction is performed on the filtered signal i.e. within the receiver passband.

NR (NOISE REDUCTION)

The noise reduction system in the IC-7300 is very effective. Pressing the NR button turns the noise reduction on for the active receiver. Holding the NR button down brings up a sub-menu where you can adjust the noise reduction level (default 5). Adjust the level to a point where the noise reduction is effective without affecting the wanted signal quality.

Noise reduction filters are aimed at wideband noise especially on the low bands rather than impulse noise which is managed by the noise blanker. The NR filter works best when the received signals have a good signal to noise ratio. By introducing a very small delay, the DSP noise reduction filter is able to look ahead and modify the digital data streams to remove noise and interference before you hear it.

AF, RF/SQL

The black inner knob controls the AF Gain (volume).

The silver outer knob can be configured in three ways. In the normal mode turning the knob clockwise from the '12 o'clock position' sets the squelch. Adjust it so that the receiver audio is muted when the receiver is only receiving noise. The green RX LED will go out. When a signal is received the squelch will open and you will hear the transmission. Turning the knob anti-clockwise from the '12 o'clock position' reduces the RF Gain of the receiver.

The squelch level is marked by a small white triangle at the top of the S meter. A green LED to the right of the display indicates that the receiver squelch is open. While transmitting the LED turns red.

To set the RF/Squelch mode select, <MENU> <SET> <Function> <RF/SQL Control>. The options are; AUTO, SQL, or RF+SQL. The default setting is RF+SQL. It works as described above. The AUTO mode acts as a squelch control for FM and AM operation and as an RF Gain control for SSB and the other modes. In SQL mode the control works as a squelch control only and RF Gain is set to maximum.

SD CARD SLOT

You can use a 'full size' SD card to save data settings, including; audio recorded off the receiver, voice keyer messages, screenshots of the display, logs, the RTTY decode history, and the radio's current settings. It is also be used for firmware updates. Note that the CW and RTTY keyer messages and the memory slots are stored in the radio, not on the SD card.

It can be a 2 Gb SD card or any SDHC card from 4 Gb up to 32 Gb. Icom recommends using SanDisk© SD cards. I am using a 16 Gb SanDisk SDHC card. The information screen says that after recording all of the keyer macros and saving the settings a few times that I have 257 hours of recording time left. So, buying a 16 Gb card as I did is probably overkill. I leave the SD card in the radio all the time.

Icom recommends formatting the SD card or USB stick, using the function in the radio, before using it for storage. <MENU> <SET> <SD Card> <Format>.

There is also a menu option to 'unmount' the SD card or USB stick before you remove it. (The same as you would when removing a USB stick from your PC). <MENU> <SET> <SD Card> <Unmount>.

The blue SD icon at the top right of the display indicates that an SD card is installed. It flashes when the card is being written to or read from, rather like the hard drive LED on a PC.

MAIN DISPLAY SCREEN

See 'The touch screen' in the next section, on page 60.

MULTI

Pressing the MULTI button turns on a selection of adjustments that are relative to the mode that you are using. Touch the Soft Key icon to select the item that you want to change. Turning the knob changes the value of the setting. The full set of MULTI menu controls is covered under 'MULTI Soft Keys' on page 71.

Touching a Soft Key on the MULTI menu often turns a function on or off. If a function is turned on, a blue indicator appears to the left of the control. If a second touch creates a different setting, an orange indicator may replace the blue one. Most Soft Keys that have multiple options don't have the blue or amber indicator. For example, the Notch Filter MULTI menu Soft Key cycles through the three manual notch width options; NAR, MID, and WIDE. If you touch the POSITION Soft Key, you can use the Multi knob to set the notch frequency within the audio passband.

Pressing the MULTI button again turns off the selection screen. So, does touching the main display, pressing the EXIT button or the Return icon ↰.

➢ **Allocating a custom function to the Multi knob**

Turning the MULTI knob when there the Multi menu is not visible, usually adjusts the VFO in 1 kHz steps, or it steps through the channels in memory channel mode.

The selected option is indicated with a white icon in the top right of the display next to the clock. Pressing the RIT or ◢TX buttons allocates the offset adjustment to the MULTI knob. You can also make the MULTI knob perform other options. If you touch and hold an icon while it is open in the MULTI menu or a popup, will allocate that function to the Multi knob until you press and hold the knob to clear it. The custom options are indicated with an orange icon in place of the white one.

Function	Action	Control	Icon
Clear	Press and hold	MULTI	White
kHz	Turn	MULTI in VFO mode	White
M-CH	Turn	MULTI in memory mode	White
RIT	Press	RIT button	White
◢TX	Press	◢TX button	White
RIT◢TX	Press	RIT◢TX buttons	White
RF PWR	Touch and hold	RF POWER Multi menu	Orange
MIC G	Touch and hold	MIC GAIN Multi menu	Orange
COMP	Touch and hold	COMP Multi menu	Orange
MONI	Touch and hold	MONITOR Multi menu	Orange
SPEED	Touch and hold	KEY SPEED Multi menu	Orange
PITCH	Touch and hold	CW PITCH Multi menu	Orange
NB LVL	Touch and hold	NB LVL Popup menu	Orange
NB DEP	Touch and hold	NB DEPTH Popup menu	Orange
NB WID	Touch and hold	NB WIDTH Popup menu	Orange
NR LEV	Touch and hold	NR LEVEL Popup menu	Orange
NOTCH	Touch and hold	NOTCH Popup menu	Orange
VOX G	Touch and hold	VOX GAIN Popup menu	Orange
A-VOX	Touch and hold	ANTI-VOX Popup menu	Orange
VOX D	Touch and hold	VOX DELAY Popup menu	Orange
BKIN D	Touch and hold	BKIN DELAY Popup	Orange

XFC

XFC stands for 'transmit frequency check.' It is quite a useful feature. It lets you check and listen to the frequency that you will transmit on. In the normal non-split mode pressing XFC opens the receiver squelch so that you can hear the frequency that the Main VFO is set to. In the split mode, pressing XFC opens the receiver squelch on the transmit (split) frequency. It also changes the split frequency display to show the split offset in kHz. Hold down XFC and turn the VFO knob to change the split offset.

If you are using RIT or ◢TX (receive or transmit incremental tuning) pressing XFC will show you the actual frequency that you will transmit on. In the CENT screen mode, the whole panadapter will shift so that the transmit frequency is in the centre and the yellow T marker will be displayed. In the FIX screen mode, the yellow T marker will be displayed at the transmit frequency.

TX/RX LED

A green light indicates that the radio is receiving, and the squelch is open.

No light indicates that the radio is receiving, and the audio is squelched.

A red light indicates that the radio is transmitting.

AUTO TUNE

You might think that the AUTO TUNE button has something to do with the antenna tuner, but it does not. It operates in CW mode to pull the receiver onto the correct frequency while receiving a CW signal. It adjusts the VFO frequency until the received CW signal is at the tone set by the PITCH control. At that point, the received CW signal is exactly netted with the transmit frequency.

SPEECH / LOCK

Press SPEECH ⌨ to activate the Speech function. A female voice tells you the current S meter reading [optional], frequency digits, and mode.

There are a range of options under <MENU> <SET> <Function> <SPEECH>. You can choose from English or Japanese language. Other languages may possibly be available in other countries. You can set the speed of the delivery. You can set whether the message includes the S meter reading or not, and you can set the volume of the announcement. The MODE SPEECH setting selects whether you want an automatic announcement every time you change modes.

Press and hold SPEECH ⌨ to lock the VFO. This locks the main tuning knob and stops the radio jumping to another frequency if you accidentally bump the main tuning knob. This function may be useful if you are operating as a 'Run' station in a contest or you are a DXpedition station or a Net controller.

If the <Lock Function> is set to the MAIN DIAL, the rest of the touch screen is unaffected. You can still change the focus of the VFO from A to B and operate the V/M button although the frequency remains locked and you can activate split. During split operation, the Split Lock function can be used. Select <MENU> <SET> <Function> <SPLIT> <SPLIT LOCK> ON.

With the Split Lock set to ON, you can adjust the split transmit frequency in the usual way by holding down the XFC button and tuning the main VFO knob. With Split Lock turned off you can't adjust the transmit frequency in this way while the VFO is locked.

If <Lock Function> is set to PANEL, the whole display screen is locked as well as the main VFO. The panadapter and spectrum display stay working but the VFO switch V/M and split are disabled.

You can reverse the Speech and Lock functions so that Speech becomes the 'press and hold' function, and Lock becomes the 'press' function. Select <MENU> <SET> <Function> [SPEECH/LOCK] Switch and change from SPEECH/LOCK to LOCK/SPEECH.

RIT

RIT is 'receive incremental tuning.' It shifts the Main receiver frequency by up to ± 9.99 kHz without changing the VFO frequency or therefore your transmit frequency. It is usually used to tune in a station that is transmitting a little off frequency, but it can be used as a method of operating split. The amount of RIT offset is shown on the right of the screen at the top of the panadapter display. Use the MULTI knob to adjust the RIT offset. Press and hold RIT to clear the RIT Offset. Press RIT again to turn off the RIT function.

If the panadapter is in FIX mode, the green R marker shows the actual receive frequency including the RIT offset. Pressing XFC shows the T marker indicating the transmit frequency and it also lets you listen to the frequency that you will transmit on.

If the panadapter is in CENT mode, the line at the centre of the panadapter shows the actual receive frequency including the RIT offset. The whole panadapter jumps to the transmit frequency when you press XFC or when you transmit.

◢TX

Delta Transmit, often called XIT, offsets the transmitter from the displayed VFO frequency. Pressing XFC or displaying the T marker using the MARKER Soft Key shows the transmit frequency and it also lets you listen to the frequency that you will transmit on. Use the MULTI knob to adjust the ◢TX offset.

Press and hold ◢TX to clear the transmit offset. Press ◢TX again to turn off the function.

RIT AND ◢TX

Selecting both RIT and ◢TX offsets both the receiver and the transmitter to the same offset from the displayed VFO frequency. I am not sure why you would want to do that. But you can.

CLEAR

This one is a bit strange. It sets the RIT and ◢TX offset back to zero. Which is exactly the same as holding down either the RIT or the ◢TX button. It seems to me that this button is rather redundant and would have been better used for something else.

If menu setting <MENU> <SET> <Function> <Quick RIT/ ◢TX Clear is set to ON. Pressing CLEAR sets, the RIT or ◢TX offset back to zero.

If menu setting <MENU> <SET> <Function> <Quick RIT/ ◢TX Clear is set to OFF. Pressing CLEAR does nothing. You have to hold it down to set, the RIT or ◢TX offset back to zero.

SPLIT

Pressing SPLIT engages split mode. Split operation in the IC-7300 is different from many transceivers where a pre-defined 5 kHz or similar offset is applied. In the IC-7300 split means that the transceiver will transmit on the frequency indicated by the Sub VFO display, usually VFO-B. That might be set to a completely different band and mode. Split operation is indicated with an orange 'SPLIT' indication on the right of the display below the clock.

Holding SPLIT down can have two different effects depending on the split menu setting. Normally it sets the sub VFO (VFO B) to the same frequency and mode as VFO A. But, if QUICK SPLIT has been set to OFF, holding down the SPLIT button has no effect.

To change the split settings, select; <MENU> <SET> <Function> <SPLIT>.

'Split' can be used in FM mode to set the offset for a repeater. You can set a standard offset for the HF bands and a separate repeater offset for the 6m band. Using split on FM will turn on a CTCSS tone if one has been programmed. See page 35.

For more information on how to use the split mode effectively, see 'Operating split' in the chapter on 'Operating the radio,' (page 75).

A/B (VFO A TO VFO B)

The IC-7300 has two VFOs. The A/B switch selects which one is currently active. The other VFO will be used as the transmit frequency if you press SPLIT.

Using the second VFO can be useful if you are waiting for someone to come up on a frequency. Or you could leave the second VFO on a frequency that you use a lot. You can quickly flick over to the other frequency to see and hear what's happening. Touch and holding A/B sets both VFOs to the currently displayed frequency.

V/M (VFO TO MEMORY MODE)

The V/M button changes the radio from VFO mode to memory channel mode. You can either change to memory mode and then step through the channels using the up and down buttons or you can step to the channel that you want and then press V/M. In memory mode, the VFO A or VFO B indicator is replaced by MEMO. The touch and hold function sets the VFO to the frequency that is stored in the memory slot.

▼▲ BUTTONS

I don't know why the up and down buttons are not used to change bands when the radio is in VFO mode. But they aren't. They step through the memory channels. This will not change the frequency of the radio unless it has been set to memory mode by pressing the V/M button. You can either change to memory mode and then step through the channels or you can step to the channel that you want and then press V/M. Touch and hold scrolls through the memory slots fast.

These buttons can now be customised to open the preset screen, or to send a voice or keyer message. But not to act as band up and down buttons.

MPAD

Pressing the MPAD button sets the VFO to the last stored memory pad frequency and mode. Pressing repeatedly cycles through the five (or ten) stored frequencies.

Press and hold the MPAD 'memory pad' button to save the currently active VFO frequency, mode etc. into the memory pad. This is pretty handy if you want to remember a frequency and come back to it later, perhaps when the current QSO on the frequency has concluded.

The memory pad can store either 5 or 10 frequencies, depending on the setting, <MENU> <SET> <Function> <Memo Pad Quantity> on page five of the function submenu. The memory stores the mode, frequency, filter setting, CW break-in, VOX, compressor, and AGC settings, but not the split offset.

Saving the active VFO frequency into a memory pad position moves the other frequencies down in the list. The previous contents of the bottom memory slot are discarded.

MAIN VFO KNOB

This is the main tuning knob. You know what it does. It is also used to set some sub-menu controls such as the filter bandwidth adjustment I like the tuning knob it has a nice weight. Under the dial, there is a lever to adjust the drag on the knob. I have never felt a need to adjust it.

The touch screen display with many options selected

CUSTOMISED BUTTONS

You can change the functions of four of the front panel controls. The VOX/BK-IN button, the AUTOTUNE button, and the Up and Down ▼▲ buttons.

Use <SET> <Function> <Front Key Customize> to change the functions.

Any of the four buttons can be changed from their default to,

1. Open the PRESET screen

2. Send voice/CW/RTTY message 1 (dependent on the mode)

3. Send voice/CW/RTTY message 2 (dependent on the mode)

4. Send voice/CW/RTTY message 3 (dependent on the mode)

5. Send voice/CW/RTTY message 4 (dependent on the mode)

Weirdly you can also reverse the Memory Channel up and down buttons, but you can't use them for band-up and band-down.

The touch screen

The touch screen display is probably the most important feature of the radio. It is used to provide controls that are not available as knobs or buttons on the front panel, access to the many menus and settings, the display of the operating parameters and of course the for the fabulous panadapter and waterfall display.

Having a high-resolution display allows the menu items to be presented in plain English rather than the cryptic codes and number systems used in older transceivers. This makes them much easier easy to understand. But there are a lot of settings available and they are spread over five screens. Which makes it hard to remember where to find particular settings. For example, is the video bandwidth setting under <Menu> <Set> <Display> or the EXPD/SET menu? Different menu functions are accessed from the MENU, FUNCTION, QUICK, and MULTI buttons and the EXP/SET 'Soft Key' icon.

According to the Icom manual, there are 25 different indicators on the LCD display. I won't bore you by listing each one. Instead, I will mention them as part of my explanation of the radio's functions. The full list is on the Icom Basic manual pages 1-4 and 1-5.

The following items are adjustable by touching the icons on the touch screen.

TIME

The time is displayed in the top right of the screen. Touching the time display opens a window displaying the local time and date and UTC time and date. The window stays open until you Press EXIT, or touch the Return icon ↺ to exit.

METER DISPLAYS

Touch the meter scale to cycle through the transmitter meter options.

The options are Po (power out), SWR (standing wave ratio), ALC (automatic level control), COMP (audio compression), V_D (power amp FET drain voltage), and I_D (power amp FET drain current).

On the expanded display with the larger panadapter and on the display with the small panadapter and no on-screen menu icons, the meter is rather small. But it works the same way.

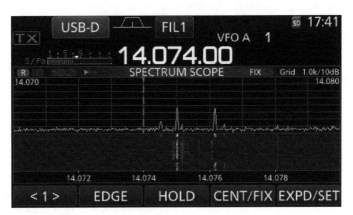

Expanded spectrum display

While the radio is receiving, the meter always reads the received signal strength in S points. The squelch level is indicated by a small white triangle on the top of the S meter scale.

> **The multi-function meter**

Touch and hold the meter scale or press the MENU button and touch the METER icon to display the multi-function meter. Touch and the multi-function meter again to close it. Or use the EXIT button.

The multi-function meter displays all of the transmit metering at the same time and also an indication of the temperature measured at the final transistors. This display is ideal when you are setting up the transmit levels for SSB because you can monitor the power output, ALC reading, and compression at the same time.

MODE

Touching the blue 'mode' icon brings up the MODE selection menu.

- SSB selects the correct SSB mode for the current band (USB or LSB) then exits the menu screen.
    - o   If the radio is already on USB, touching SSB will change the radio to LSB mode.
    - o   If the radio is already on LSB, touching SSB will change the radio to USB mode.
- CW selects the CW mode.
    - o   If the radio is already on CW it will select CW-R. This reverses the beat frequency oscillator so that it is lower than the received CW signal (CW-USB mode), rather than the default (CW-LSB mode) where the beat frequency oscillator is higher than the received CW signal (CW-LSB mode).
    - o   If the radio is already on CW-R it will return the radio to CW.
- RTTY selects the built-in RTTY mode.
    - o   If the radio is already on RTTY it will select RTTY-R. This changes the operating sideband and more importantly it reverses the Mark and Space frequencies. If decode is gibberish the other station might be transmitting on the other sideband, creating RTTY-R.
    - o   If the radio is already on RTTY-R it will return the radio to RTTY.
- AM selects AM mode
- FM selects FM mode
- DATA selects the Data mode where the transmit modulation is supplied over the USB cable from an external program, or over the ACC jack from an external device such as a Keyer, TNC, or PC.

    There are four data modes. The one that you are most likely to use is USB-D, the SSB data mode. You might possibly use FM-D for FM packet radio.

*Note that audio from the receiver is sent to the PC over the USB cable in all modes. But audio from the PC will only modulate the transmitter if the radio is in a **data mode.***

CW can be operated from a PC while using the CW mode as it uses the RTS or DTR line for signalling rather than an audio tone. FSK (not AFSK) RTTY can be operated from a PC while using the RTTY mode as it uses the RTS or DTR line for data signalling rather than an audio tone.

- There are four data modes. USB-D, LSB-D, AM-D, and FM-D
- Note to return to SSB from USB-D you have to touch DATA again rather than SSB. "Strange but true!" Touching SSB will not return you to USB. It will switch to LSB mode. You have to get into the mindset that the Data modes are subsets of the SSB, AM and FM modes. From SSB, touch data to get to USB-D and touch data again to exit back to USB.

FREQUENCY DISPLAY

The frequency display has several touch screen functions.

➢ **MHz digits**

Touch the MHz digit(s) to bring up the band change menu. There are three options.

- Touch the band you want to change to.
- Touch and hold a band icon, to bring up the band stacking register.
- To enter a frequency directly into the VFO, set a split offset, or select a memory channel touch F-INP.
  - ○ CE clears an entry.
  - ○ ENT enters the data and closes the screen. It shortcuts the entry process. For example, type <1> <4> <ENT> and the radio will set 14 MHz into the VFO.
  - ○ Touch a digit and then split to enter a split, e.g. <0> <5> <Split> sets a –5kHz split.
  - ○ If the radio is in memory mode, rather than VFO mode you can touch a number then MEMO to change to that memory channel. For example, <2> <1> <MEMO> changes to memory channel 21. Unfortunately, there is no indication of what is in any particular memory slot, so it is a bit "hit and miss."
  - ○ Return ↺ closes the screen.

Touch and hold the MHz digit(s) to bring up the band stacking register for the current band. Then select one of the three displayed frequencies.

➢ **kHz digits**

Touch the three kHz digits to toggle the VFO tuning step between 'fast' and 'normal.' The 'normal' tuning step is either 10 Hz per step or 1 Hz per step, (see Hz digits below). A white triangle above the 1 kHz digit indicates that the VFO is in fast tuning mode. Touch and hold the three kHz digits to change the 'fast' tuning step (TS). You can set different 'fast' tuning steps for each mode. The default for SSB is 1 kHz but, I prefer 0.1 kHz (100 Hz).

➢ **Hz digits**

Touch and hold the Hz digits to toggle between 'normal' 10 Hz per step tuning and 1 Hz per step tuning.

➢ **Tuning rate**

The tuning rate may speed up as you turn the main VFO knob. This is normal and is linked to the speed that you turn the knob. The feature is designed to get you to the other end of the band quickly when required. If you don't like it, you can turn it to High, Low, or Off using <MENU> <SET> <Function> <MAIN DIAL Auto TS>.

➢ **¼ tuning speed**

The digital modes and CW allow the use of the ¼ tuning function. Select <FUNCTION> <1/4> to turn the function on or off. The function slows down the tuning rate of the VFO to make tuning in narrow CW and digital mode signals easier. It is indicated with a ¼ icon to the right of the 10 Hz digit of the frequency display.

VFO mode display: In VFO mode, the VFO frequency is displayed in large white letters. The currently selected memory position is displayed below the VFO A or VFO B indication on the right. If the memory position is unassigned the word Blank is placed under the memory position number. Rather strangely this space is not used to display the name of assigned memory slots. A number preceded by a star in this display position indicates that the channel is in a scan group. Pressing the ▼Down and ▲Up buttons changes the memory slot, but this has no effect on the radio as it is in VFO mode. I don't know why Icom didn't use these two buttons for Band up and down switches while the radio is in VFO mode.

VFO mode   **MEMO memory channel mode**

Memory mode display: In memory mode, the line indicating the current VFO changes to MEMO. The memory slot number and scan group (or Blank) are displayed below. In the large digit display, the name of the memory channel is displayed above the 10s digit of the frequency display. Press the ▼Down and ▲Up buttons to change the memory slot.

SPECTRUM AND WATERFALL DISPLAY

Touching the screen within the spectrum scope/waterfall area expands a section of the display. Touching within the square moves the VFO to a frequency within the box. You will probably still have to use the main tuning knob to get the signal exactly tuned in. Touching the display anywhere outside of the box exits the expanded display with no effect on the VFO tuning.

The spectrum scope/waterfall display Soft Keys are only available with the medium or large panadapter displays. They are not available with the small panadapter display. This is a little annoying if you have the Voice message, Keyer, or Decoder screens open. You have to make sure that the panadapter is exactly how you want it before opening any of those other screens.

CENTER AND FIXED SPECTRUM DISPLAYS

The CENT/FIX Soft Key switches the spectrum display between the CENTER and FIXED spectrum displays. The 'fixed' display mode is better suited to observing a nominated band segment for activity. The 'Center' display mode is better suited to tuning across a band.

The February 2021 (V1.40) firmware update introduced a scroll mode to the FIX and CENT spectrum display modes. This is a good idea because it stops you tuning off the edge of the displayed spectrum. Touch and hold the CENT/FIX Soft Key to enter the SCROLL-C or SCROLL-F scrolling mode.

PANADAPTER BAND EDGES

When the panadapter is set to the FIX display, touching the EDGE icon on the expanded display cycles through four panadapter bandwidth settings.

Note that the panadapter band edges are not the same as the amateur radio band edges that set the frequency ranges that you can transmit on and the band edge beeps that tell you when you have tuned outside of an amateur radio band. See page 43.

You can customize the panadapter band segments. Thirteen frequency zones allow four 'Fixed Edges' (panadapter bandwidths) each. For example, on the 20m amateur band, I set the first option to display the whole band from 14.000 to 14.350 MHz. The second option displays the CW and digital part of the band, 14.000 to

14.100 MHz and the third option displays the SSB band segment from 14.100 to 14.350 MHz. You can set any band edges you like, up to the maximum panadapter bandwidth of 1 MHz.

➤ Setting the panadapter band edges

If the Soft Key icons are not currently displayed at the bottom of the touch screen, hold <M.SCOPE> down for one second to enable the expanded screen.

Touch and hold <EXPD/SET> to open the 'Scope Set' menu and then select <Fixed Edges> at the bottom of the fourth setup screen.

Touch one of the thirteen frequency ranges and then touch one of the four band edges. Set the lower frequency then press <ENT> and then set the upper edge frequency and touch <ENT>. The minimum panadapter bandwidth you can set is 5 kHz. The maximum is 1 MHz.

Touch and hold any of the four band edges to reset it to the Icom default.

➤ Scroll-F display mode

If you touch and hold the CENT/FIX Soft Key while the radio is in the FIX display mode, the FIX icon at the top of the spectrum display will change to SCROLL-F on a green background. Now, when you tune outside of the fixed frequency range the display will jump to the next band segment the same size as the fixed edge, and the receiver cursor will jump to the start of that segment, (tuning up or down). Selecting EDGE will change the display to a scan width equal to the width of the next fixed edge, but it will not change the VFO to the frequencies programmed into that fixed edge, i.e. the band edge segment sizes become the span size for the display.

➤ CENTER display mode

If necessary, touch the CENT/FIX Soft Key to change to the CENTER display. The spectrum and waterfall display act like a band-scope, displaying signals below and above the VFO frequency. The frequency that the receiver is tuned to is always in the centre of the spectrum display. Touching SPAN changes the spectrum bandwidth.

➤ Scroll-C display mode

If you touch and hold the CENT/FIX Soft Key while the radio is in the CENT display mode, the CENT icon at the top of the spectrum display will change to SCROLL-C on a brown background. This switches the display to a FIX(ed) display with the span size set by the SPAN setting.

When you tune outside of the fixed frequency range the display will jump to the next band segment the same size as the Span setting and the receiver cursor will jump to the start of that segment, (tuning up or down). Selecting SPAN cycles through the span settings from ±2.5 kHz to ±500 kHz.

SPECTRUM SCOPE SOFT KEYS

The spectrum scope Soft Keys are displayed at the bottom of the touch screen. If they are not visible press and hold the M.SCOPE button.

The touch screen Soft Keys

> **< 1 >**

Change to Menu 2 selections.

> **SPAN**

When the panadapter is set to the CENT display, touching the SPAN icon on the expanded display cycles through eight panadapter bandwidth settings.

±2.5 kHz, ±5 kHz, ±10 kHz, ±25 kHz, ±50 kHz, ±100 kHz, ±250 kHz, and ±500 kHz.

Touch and holding SPAN sets the panadapter bandwidth to ±2.5 kHz.

> **EDGE**

In the FIX spectrum mode, EDGE cycles through four pre-set band edges per frequency zone.

➢ **HOLD**

Freezes the spectrum scope and waterfall.

➢ **CENT/FIX**

Toggles between the FIX and CENTER span modes. Touch and hold to enter the Scroll-F and Scroll-C modes.

- FIX mode displays the spectrum between pre-defined band edges. See EDGE.

- CENT displays the spectrum below and above the active VFO frequency. See SPAN.

- SCROLL-F mode displays the spectrum with a display span (width) defined by the pre-defined band edges. When you tune outside of the displayed bandwidth the display scrolls to the next band segment.

- SCROLL-C mode displays the spectrum with a display span (width) defined by the SCAN bandwidth. When you tune outside of the displayed bandwidth the display scrolls to the next band segment.

➢ **EXPD/SET**

Touch EXPD/SET to turn on or off the expanded spectrum scope display. Touch and hold EXPD/SET to access the SET menu.

➢ **< 2 >**

Change to Menu 1 selections.

➢ **REF**

Adjustment of the Spectrum display reference level. Annoyingly there is only one reference level control, so you often have to change the REF level if you change bands and sometimes if you move from a narrow panadapter span to a wide one.

➢ **DEF**

Returns the Spectrum display reference level to 0 dB.

➢ **SPEED**

Cycle through FAST, MID, SLOW panadapter and waterfall speeds. The speed is indicated with blue markers on the top line of the Spectrum Scope.

➢ **MARKER**

Displays the orange (T) transmit frequency marker. In the FIX spectrum mode, the green (R) receive frequency marker is always displayed.

The full table of settings is in the SCOPE Spectrum Scope Soft Keys section on page 89. But there are a couple of scope settings that I believe most users will want to change immediately.

➢ **CENTRE Type Display**

This is a very important setting. It sets the way that the wanted signal is displayed on the panadapter display. The Icom default setting is rather odd, placing the panadapter centre in the middle of the filter passband instead of at the carrier point. I believe that most users will want to select either the Carrier Point Center (Abs Freq) option or the Carrier Point Center. Touch and hold <EXPD/SET> <CENTER Type Display> and set it to <Carrier Point Center (Abs Freq)>. See page 92.

➢ **Marker Position (FIX Type)**

This is a very important setting. It sets the way that the wanted signal is displayed on the FIX mode panadapter display. The Icom default setting is rather odd, placing the panadapter centre in the middle of the filter passband instead of at the carrier point. I believe that most users will want to select the 'Carrier Point' option. Touch and hold <EXPD/SET> <Marker Position (FIX Type)> and set it to <Carrier Point>. See page 92.

SPECTRUM SCOPE SET MENU

The full table of settings is in the SCOPE Spectrum Scope EXPD/SET section on page 92. I changed the colour of the spectrum trace to make it look like a spectrum analyser line display and I turned off the peak hold display. I also changed the 'Waterfall Size (Expand Screen)' setting which lets you set how much waterfall is displayed on the expanded spectrum scope display.

SETTING THE SPECTRUM VIEW TO A 'LINE' VIEW

I don't like the default spectrum display, which is a filled spectrum with peak hold. It's a personal preference, you might like it. I guess that I am used to spectrum analysers and SDR receivers which have a line display rather than a filled in trace. Although there is no option for a line display there is a 'workaround' that works for me.

If the Soft Key icons are not currently displayed at the bottom of the touch screen, hold <M.SCOPE> down for one second to enable the expanded screen.

Touch and hold <EXPD/SET> to open the 'Scope Set' menu. On the SCOPE SET menu, make the following changes.

<Max Hold>	OFF
<Waveform type>	Fill+Line
<Waveform Color (Current)>	R:0 G:0 B:57
<Waveform Color (Line)>	R:200 G:200 B:200

If you don't like the changes you can set the display lines back to default by touch and holding the menu item and selecting the 'Default' option.

MEMORY MANAGER

On the screens where the VFO frequency is large and the VFO and memory number are on the right side of the display, you can access the memory management screen by touching the memory number or the text that says 'VFO' or 'MEMO'.

This screen offers a completely different way to manage the radio memories than using the screen that is accessed by pressing <MENU> <MEMORY>.

VFO sets the VFO mode and MEMORY sets the memory mode, exactly the same as pressing the V/M button. In the VFO mode, MW saves the current VFO frequency, mode etc. to the current memory slot as indicated beside the VFO A or B indicator. M-CLR clears the current memory slot. SELECT does not, as you might expect, select the memory and pop it into the VFO. Instead, if the memory mode is selected and the current memory slot is not blank, you can use SELECT to change the memory slot, scan group.

FILTER

The current filter settings are indicated by the filter icon to the right of the mode indicator at the top of the display. A tiny white dot to the right of the icon indicates that Twin PBT (twin passband tuning) is active. Touch FIL to cycle through the Filter 1, 2, and 3 bandwidths for the currently selected radio mode.

Touch and hold FIL to open the filter setup screen. Here you can see the effects of any Twin PBT adjustments, and you can set the filter bandwidth with a sharp or soft roll-off (for BW >500 Hz only).

To adjust the bandwidth, select FIL1, FIL2, or FIL3 then touch the BW Soft Key. Adjust the bandwidth using the main VFO knob.

MULTI SOFT KEYS

Pressing MULTI opens a set of Soft Keys on the right side of the display. The Soft Key icons are specific to the operating mode or function that you have selected. You can select the items by touching the appropriate icon and adjust them by turning the MULTI control knob.

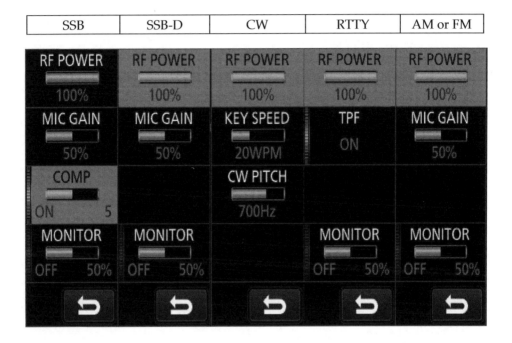

SSB	SSB-D	CW	RTTY	AM or FM
RF POWER 100%	RF POWER 100%	RF POWER 100%	RF POWER 100%	RF POWER 100%
MIC GAIN 50%	MIC GAIN 50%	KEY SPEED 20WPM	TPF ON	MIC GAIN 50%
COMP ON 5		CW PITCH 700Hz		
MONITOR OFF 50%	MONITOR OFF 50%		MONITOR OFF 50%	MONITOR OFF 50%
⮌	⮌	⮌	⮌	⮌

The COMP, MONITOR, and TPF settings with 'on/off' features in addition to an adjustable control have a blue indicator at the left of the selection to indicate that the function is on.

Touching the screen anywhere outside of the menu area, touching the Return ⮌ icon, pressing MULTI again, or pressing EXIT will close the menu selection.

➢ **RF POWER**

Sets the transmitter's output power. Press the MULTI button and touch the RF Power icon. Set the power for 100% by turning the Multi knob. Although Icom makes no recommendation regarding transmitter power on digital modes, you might want to reduce the power a little if you are transmitting 'continuous duty' modes like PSK or RTTY for extended periods. I run full power for FT8. When using a linear amplifier, use this control to reduce the RF Power to a level that won't overdrive the amplifier.

➢ **MIC GAIN**

The Mic Gain control is used to set the modulation level on voice modes. Turn the MULTI knob to change the setting. I adjust it on SSB mode with the compressor turned off. See 'Setting up the radio for SSB operation' on page 9. The Mic Gain should only have to be adjusted once and the adjustment made in SSB mode should be OK for FM and AM as well as for SSB.

➢ **COMP level**

Touch to turn on the compressor. A blue indicator bar and the word ON indicate that the compressor is active. Also, COMP is displayed under the MHz digit of the VFO display. The default level is 5 and that level works for me. Turn the MULTI knob to change the setting. See 'Setting up the radio for SSB operation' on page 9

➢ **MONITOR**

The Monitor Soft Key turns on the audio monitor so that you can hear the signal that you are transmitting. Turning the Multi knob adjusts the level of the monitor signal. 'Monitor' is not available in the CW mode because sidetone is always on.

A blue indicator and the word ON indicate that the transmit monitor is turned on.

Unless you specifically want to listen to your transmit signal it is less distracting to leave the Monitor turned off. The SSB voice keyer will be heard if the 'Auto Monitor' setting is 'on,' irrespective of the Monitor setting.

➢ **KEY SPEED**

When the radio is in CW mode the CW speed is adjustable from 6 wpm to 48 wpm. Turn the MULTI knob to change the setting.

➢ **CW PITCH**

The CW pitch is adjustable from 300 Hz to 900 Hz. It is a personal preference. I use 700 Hz. Turn the MULTI knob to change the setting.

➢ **TPF (Twin Peak Filter)**

The TPF (Twin Peak Filter) filter is available in the RTTY mode. It is very effective at lifting weak RTTY signals out of the noise. When TPF is turned on, a blue bar is displayed to the left of the control.

The TPF has a peak at the RTTY Mark and Space audio frequencies, emphasising the RTYY signal and reducing noise elsewhere in the audio spectrum. TPF can make tuning an RTTTY signal a bit tricky, so it may be easier to turn it on after you have tuned to the RTTY signal.

Some of the MULTI menus are activated by holding down front panel buttons or Soft Keys on the FUNCTION display.

CW Break-in	VOX	Manual Notch	Noise Reduction	Noise Blanker
BKIN DELAY 7.5d	VOX GAIN 50%	NOTCH POSITION	NR LEVEL 4	NB LEVEL 50%
	ANTI VOX 50%	WIDTH MID		DEPTH 8
	DELAY 0.2s			WIDTH 50
	VOICE DELAY OFF			
⤺	⤺	⤺	⤺	⤺

> **CW Break-in delay**

The CW Break-in delay submenu is activated in CW mode by pressing the FUNCTION button and then touch and holding the BKIN Soft Key or by holding down the VOX/BK-IN button. The delay setting affects the semi Break-in 'BKIN' mode. In semi break-in mode the Morse key or paddle will key the transmitter while the CW is being sent and the radio will return to receive after a delay when the key is released. The default delay is 7.5 dits at the selected key speed. In the full break-in mode, the transmitter returns to receive immediately after the key is released.

> **VOX settings**

The VOX submenu is activated in SSB, AM, or FM modes by pressing the FUNCTION button and then touch and holding the VOX Soft Key. Or by holding down the VOX/BK-IN button. The settings for VOX Gain, Anti-VOX, Delay, and Voice Delay are covered in the FUNCTION menu section, on page 125.

> **Manual Notch**

The Notch menu is accessed by holding down the NOTCH button on the front panel or by touch and holding the NOTCH Soft Key on the FUNCTION menu. The manual notch is very effective at removing annoying interference. I sometimes use it in Narrow mode to eliminate very strong FT8 signals that are affecting the other signals on the FT8 waterfall. You can touch the position icon and turn the MULTI knob to set the audio frequency in the receiver passband that the notch will attenuate. You will see this as a black zone on the WSJT-X waterfall. The second icon allows you to set a wide, medium, or narrow notch. Start with narrow and if you still hear interference consider using the wider options.

You can see the effect of the notch filter by opening the Audio Scope. Select <MENU> <AUDIO>. Tune to a frequency where you can hear a carrier 'birdie.' You must have the receiver squelch open to see signals on the audio spectrum scope. Turn on the auto notch and you will see the carrier signal disappear from the audio spectrum display and you won't be able to hear it anymore. Enable the manual notch and you will be able to see a black zone on the audio spectrum indicating a deep null in the signal. Changing from narrow to mid or wide makes the null zone wider. Adjusting the manual notch position moves the nulled band across the audio spectrum.

There are two menu settings that can affect Notch operation. For AM and SSB you can select Auto Notch only, Manual Notch only, or the default choice of both Auto and Manual notch options. Since the auto notch is best for interfering carriers and the manual option allows you to place the notch where you want it on the audio passband, I can't imagine why you wouldn't want both options. But you can change it if you want to. <MENU> <SET> <Function> <[NOTCH] Switch (SSB)> or <[NOTCH] Switch (AM)>.

➤ **NR level**

The default noise reduction level is 5. Adjust the level to a point where the noise reduction is effective without affecting the wanted signal quality.

➤ **NB level, depth and width**

The noise blanker LEVEL control (default 50%) sets the audio level that the filter uses as a threshold. Most DSP noise blankers work by eliminating or modifying noise peaks that are above the average received signal level. They usually have no effect on noise pulses that are below the average speech level. Setting the NB level to an aggressive level may affect audio quality.

The DEPTH control (default 8) sets how much the noise pulse will be attenuated. Too high a setting could cause the speech to be attenuated when a noise spike is attacked by the blanker. This could cause a choppy sound to the audio.

The WIDTH control (default 50%) sets how long after the start of the pulse the output signal will remain attenuated. Set it to the minimum setting that adequately removes the interference. Very sharp short duration spikes will need less time than longer noise spikes such as lightning crashes.

Operating the radio

The Icom IC-7300 is a joy to use. I really like the feel of the main VFO tuning knob. The drag is adjustable, but I like it just the way it came from the factory. Generally, the ergonomics are good although some things take a bit of getting used to. For example, if you are using the Decode, Voice, or Keyer screens and you change mode, the screen does not follow the mode change.

VOICE KEYER

The Voice message option is only available when the active VFO is in a voice mode or a voice-data mode. Press <MENU> VOICE> and touch a pre-recorded message T1 to T8 to send the voice messages. Recording and setting up the voice messages is covered on page 95.

Touch and holding a voice message key, T1 to T8 will cause the message to repeat until you touch the T message key gain or press the PTT on the microphone.

OPERATING SPLIT

Working in the 'Split' mode is a very common requirement if you are trying to work a rare DX station or DXpedition when they have a 'pileup' of stations calling them.

To work 'Split' you set the current VFO, we will assume that is VFO A, to receive the frequency that the DX station is using. You will transmit on the frequency indicated by VFO B. VFO A and B can be reversed. It makes no difference. On bands above 10 MHz, you will transmit USB on a frequency that is a few kHz higher than the DX station. For bands lower than 10 MHz, you will transmit LSB on a frequency that is a few kHz lower than the DX station. For CW the operation is the same, but the split offset is less. Usually 1-2 kHz rather than 5-10 KHz. You can operate split on digital modes but it is less common. If you are the rare DX station or on a or DXpedition you would, of course, reverse the split.

Split operation in the IC-7300 is different from many transceivers where a pre-defined 5 kHz or similar offset is applied. In the IC-7300 using split means that the transceiver will transmit on the frequency indicated by the Sub VFO display. That might be on a completely different band and mode, so you do have to be careful. Split operation is indicated with an orange 'Split' indication on the right side of the display, underneath the clock. The VFO B (Transmit) frequency is displayed below that.

For a typical scenario where I am attempting to contact a DX station using SSB on the 20m band. I recommend the following technique.

- Use <MENU> <SET> <Function> <SPLIT> to set <Quick Split> to ON.
- Tune the Main VFO to the frequency that the DX station is using
- Press and hold the SPLIT button, to set the Sub VFO to the same frequency as the main VFO. Turn down the Main AF volume a bit so that the DX station is quiet.
- Hold down the XFC button and use the VFO knob to set the split offset.

OR

- Tune the Main VFO to the frequency that the DX station is using
- Press and hold the SPLIT button, to set the Sub VFO to the same frequency as the main VFO. Turn down the Main AF volume a bit so that the DX station is quiet.
- Use the A/B button to select VFO B and tune the receiver through the pileup of stations to find a good frequency to transmit on. Listen to the pileup and try to work out which way the DX station is moving through the pileup or simply use the panadapter find a quiet spot. Make sure that you are transmitting within the span of frequencies the DX station is listening to. For example, "5 kHz up" or "5 to 10 kHz."
- Use the A/B button to select VFO A so that you can hear the DX station again. Now you are now ready to make your call.

OR

- Tune the Main VFO to the frequency that the DX station is using
- Touch the MHz display on the VFO. Touch F-INP. Touch a number representing the split that you want, for example <5> and then touch SPLIT. For a negative split when working LSB touch the decimal point and then the split <.> <5>.

You use the same techniques when using the low bands except the pileup will be spread on frequencies below the DX station, so you will use a negative split offset and LSB. For CW operation the technique is the same except the split will be 1-2 kHz rather than 5-10 kHz.

OPERATING CW MODE

In CW mode the receiver should initially be tuned to the exact frequency of the CW station. The centre line or receive marker should be exactly aligned with the signal shown on the spectrum and waterfall display. At that point, the audio tone will be very close to the tone set by the CW PITCH control (below). Press Auto Tune while receiving the CW signal to net the receiver frequency exactly.

➢ **Break-in setting**

In CW mode the VOX/BK-IN button controls the break-in settings. They can also be changed by pressing the FUNCTION button and using the BKIN Soft Key icon.

The transceiver will not automatically transmit CW unless either semi break-in (BKIN) or full break-in (F-BKIN) has been selected. The current break-in setting is displayed on the left just above the bar meter.

If BKIN is set to OFF the transceiver can only be made to transmit by pressing the TRANSMIT button, pressing the PTT button on the microphone, sending a CI-V command, or grounding the SEND line on the ACC jack. The break-in setting affects the sending of keying macro messages and Morse Code send from a key or paddle, but not CW sent from an external computer program.

- With BKIN OFF you can practice CW by listening to the side-tone without transmitting.
- Full break-in mode F-BKIN will key the transmitter while the CW is being sent and will return to receive as soon as the key is released. This allows for 'QSK' reception between CW characters.
- Semi break-in mode BKIN will key the transmitter while the CW is being sent and will return to receive after a delay when the key is released. Touch and hold the BKIN Soft Key or hold the VOX/BK-IN button to adjust the delay. Turn the Multi knob to change the setting. The default is a period of 7.5 dits at the selected key speed.

➢ **MONI (Monitor)**

The MONI (transmit monitor) function is disabled in CW mode because the sidetone is always turned on. If you don't want sidetone you can turn down the level to zero. (See Sidetone below).

➢ **Sidetone**

The CW sidetone is 'always on' but you can set the level to zero if it is annoying, or your key, paddle or bug has its own sidetone.

In CW mode select <MENU> <KEYER> <EDIT/SET> <CW-KEY SET> <Sidetone Level>

➢ **Auto Tune**

The AUTO TUNE button pulls the receiver onto the correct frequency while receiving a CW signal. It adjusts the receive frequency until the received CW signal is at the tone set by the CW PITCH control. At that point, the received CW signal is exactly netted with the transmit frequency. A red AUTO TUNE indication is displayed. You may have to press the button a couple of times to get the setting exact.

➢ **Key speed**

While in CW mode press the MULTI button and touch the KEY SPEED icon. Turning the MULTI knob sets the speed of the electronic keyer, including CW sent from the KEYER SEND macros. The key speed is adjustable from 6 wpm to 48 wpm.

➢ **CW pitch**

The CW PITCH control on the MULTI menu changes the pitch of a received CW signal without changing the receiver frequency. You can set the control so that CW sounds right to you.

The CW PITCH frequency is depicted below the centre of the filter passband image. Touch and hold the FIL icon at the top of the screen. The centre frequency of the passband is the pitch frequency.

➢ **CW message keyer**

When in CW mode press MENU then KEYER to show the eight CW messages. The CW messages are handy for DX or Contest operation or just to save you sending the same message over and over. They are great for sending CQ on a quiet band.

Touching one of the **M1 to M8 Soft Keys** sends the CW message. You can stop it by touching the Soft Key again or by sending with the key or paddle

Touch and hold one of the **M1 to M8 Soft Keys** to keep sending the message until you stop it by touching the Soft Key again or by sending with the key or paddle.

The M2 Soft Key is setup for contesting with an incrementing number after the signal report. If you need to send the same number again, you can decrement the number with the -1 Soft Key. The radio remembers the last number sent even if you have turned it off in the meantime, so you won't lose the number during a 48 hour contest. See 'KEYER sub-screen Soft Keys' on page 97 to set up the keyer memories.

> ¼ **tuning speed**

The digital modes and CW allow the use of the ¼ tuning function. Select <FUNCTION> <1/4> to turn the function on or off. The function slows down the tuning rate of the VFO to make tuning in narrow CW and digital mode signals easier. It is indicated with a ¼ icon to the right of the 10 Hz digit of the frequency display.

OPERATING RTTY

The radio supports three kinds of RTTY operation. Firstly, there is the onboard RTTY decoder. Which can be used with the eight RTTY message memories.

In this mode, you can take advantage of the excellent TPF (twin passband filter). I recommend using the TPF filter all the time because it really helps with accurate decodes. The second method is to use external PC software such as; MixW, MMTTY, MMVARI, Fldigi etc. with AFSK (audio frequency shift keying). AFSK uses two audio frequencies to create the frequency shift keying in the SSB mode. The third method is the FSK mode which uses a digital signal to key the transceiver to predefined mark and space offsets.

> **RTTY decode**

Set RTTY mode with a narrow panadapter ±2.5 kHz or ±5 kHz. This must be done **before** you turn on the DECODE screen.

Selecting <MENU> <Decode> while in RTTY mode displays the internal RTTY decoder. The DECODE option is not visible on the main MENU unless the radio is in the RTTY mode. Selecting DECODE will shrink the panadapter display.

Tune the VFO frequency so that the panadapter centre line or receive marker is aligned with the right (higher frequency) of the two RTTY lines. Then fine tune until the two RTTY signal peaks are lined up with the vertical lines on the audio spectrum display on the DECODE screen. This is easier to do with the TPF filter off. There is also a tuning indicator at the top left of the audio spectrum display. After you have the signal tuned you can turn on the excellent TPF (twin peak filter) which is specially designed to maximise the two RTTY tone frequencies. The TPF can be turned on or off using <MULTI> <TPF>.

If you touch the TX MEM Soft Key, the RTTY message keys RT1 to RT8 are displayed. They can be triggered by touching the relevant Soft Key or from a keyboard or button pad attached to the microphone connector. After the message has been selected, the display reverts to the decode screen.

The 'Decode' screen can be made larger by touching the EXPD/SET Soft Key. This means that more decoded text can be displayed, but it overlays the panadapter display.

> ➢ **TPF filter**

Don't forget to use the fabulous TPF (Twin Peak Filter) filter it is very effective at lifting weak RTTY signals out of the noise. Tune in the signal first then turn on the filter by pressing the **MULTI** button and touching the TPF icon to enable the filter. When TPF is on, a blue bar is displayed to the left of the control.

> ➢ **¼ tuning speed**

The RTTY mode allows the use of the ¼ tuning function. Select <**FUNCTION**> <1/4> to turn the function on or off. The function slows down the tuning rate of the VFO to make tuning in narrow digital mode signals easier. It is indicated with a ¼ icon to the right of the 10 Hz digit of the frequency display.

> ➢ **AFSK RTTY from an external PC program**

To use an external PC based digital modes program for transmitting AFSK RTTY you must use the USB-D DATA mode, not the RTTY mode. First, ensure that the active VFO is in SSB mode and then touch <DATA>. The mode should change to USB-D. You can transmit RTTY on LSB-D but it is not standard even on low bands.

Audio is sent to the PC when it is in any mode, so you can use your digital mode PC software to see and decode RTTY signals. But you must be in the USB-D DATA mode to transmit AFSK RTTY from your digital mode PC software.

If you prefer to use your favourite external digital mode program to send RTTY, the easiest method is to use AFSK rather than FSK. In AFSK the RTTY signal is sent to the transceiver as audio tones rather than a digital signal. I use MixW because I like the log function and it seems to decode well. But any digital mode software should be OK. I have found that early versions of MixW will not communicate with the radio due to the Icom addressing on the USB COM port. But the latest MixW V3.1.1 and V4 versions are fine.

> ➢ **RF Power in RTTY mode**

You can run 100 Watts, but if you are prone to very long 'overs' I suggest derating the power to 75 Watts. Press **MULTI** select the RF POWER icon and reduce RF power to 75% by turning the Multi knob. Keep an eye on the temperature meter on the multi-function meter display. If the transceiver is running hot, de-rate the transmitter power.

> **FSK RTTY from an external PC program**

An advantage of using FSK rather than AFSK from an external PC program is that you use the RTTY mode on the radio rather than the USB-D data mode. That means that you can use the TX MEM messages and the TPF (twin peak filter).

Select the RTTY mode. There are no levels to set in the FSK mode. There are instructions for setting up MMTTY for FSK operation on page 30. I have not been able to get MixW to send FSK RTTY, but the AFSK mode works fine.

OPERATING SCAN MODE

Selecting <MENU> <SCAN> displays the SCAN Menu. Touching the screen, PTT from the Mic, or touching the scan button again will halt a scan that is in progress.

The radio has three scan modes. You can scan a range either side of the current VFO setting, or between the P1 and P2 memory frequencies, or you can scan through the 99 programmed memories.

- The ◢F Soft Key starts a scan from just below the current VFO frequency to just above it. You can set the scan range from ±5 kHz to ±1 MHz using the ◢F SPAN Soft Key.

- Touch PROG to scan between the P1 and P2 frequencies. PROGRAM SCAN flashes while the scan is progressing and the decimal points in the frequency display flash.

- MEMO replaces PROG if the radio is in Memory mode instead of VFO mode. It is used to scan through selected memories from the 99 stored memories. If you touch SELECT during a memory scan, the scan will restrict itself to the nominated span group. You can change the span group using the SEL No. Soft Key, but only while a scan is in progress.

Set the squelch so that the radio is squelched before using the Scan mode and the Scan will stop when a signal is encountered. You can start the scan again by touching the PROG or ◢F Soft Key. Alternatively, if Scan Resume is set to 'ON' the scan will resume after a time. If you touch the FINE Soft Key **after** a scan has started, the scan will slow down but not stop when the squelch opens. After it passes through the signal and the squelch closes again, the scan speeds up again. I like the ◢F scan mode with 'Scan Resume' set to ON.

If you have the spectrum display set to FIX you can see the M marker moving across the band as the scan progresses. If you have the spectrum display set to CENT, the whole panadapter scrolls, which plays havoc with the waterfall.

In effect, there are four scan speeds. The scan uses the VFO tuning step and the SET menu lets you select either a Fast or Slow scan speed. The fastest scan is when the VFO is in the 1 kHz step mode and the scan speed is set to fast. The slowest scan speed is when the VFO is set to 1 Hz steps and scan speed is set to slow.

Scan controls VFO mode

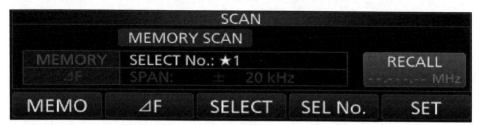

Scan controls in memory mode while scanning

Soft Key	Function	Hold	Menu Setting / Notes
PROG	Touch PROG to scan between the P1 and P2 frequencies. PROGRAM SCAN flashes while the scan is progressing and the decimal points in the frequency display flash.	None	P1 and P2 can be set in the memory screen like any other memory slot. <MENU> <MEMORY>
⊿F	Span frequencies below and above the current VFO setting. ⊿F SCAN flashes while the scan is progressing and the decimal points in the frequency display flash.	None	The range of the span is set by touching the ⊿F SPAN control.

FINE	Touch FINE while a scan is in progress and the scan rate will slow but not pause when a signal opens the receiver squelch.		
▲F SPAN	Sets the range of the ▲F scan.	None	±5 kHz, ±10 kHz, ±20 kHz, ±50 kHz, ±100 kHz, ±500 kHz, ±1 MHz.
RECALL	Touch and hold to reset the VFO to the original frequency.	VFO recall function	Touching the screen, PTT from the Mic, or touching the scan button gain will halt the scan.
SET	Sub-menu to change Scan speed and Scan resume function	None	FAST/SLOW and ON/OFF
MEMO	MEMO replaces PROG if the radio is in Memory mode instead of VFO mode. It is used to scan through selected memories from the 99 stored memories.	None	Memo steps through all the stored memories. Or through the stored memories that are tagged with any of the three scan group tags, *1, *2, *3.
SELECT	When not scanning, SELECT changes the scan group of the current memory channel. While scanning SELECT selects all channel or scan group only scan. This is buggy and sometimes it stops the scan instead. Using SEL no. then SELECT seems to fix the issue.	Clears one or all scan groups	Touch and hold clears one or all scan group settings. Be careful.
SEL No.	Selects the group of memory channels that will be scanned. *1, *2, or *3, or *1, *2, and *3.	None	SEL No. is only displayed in memory scan mode while the scan is running.

OPERATING EXTERNAL DIGITAL MODE SOFTWARE

Once the connection has been established between your digital mode software and the radio, the operation of the radio is mostly controlled from the external software. The main thing to remember is that for most modes you must be in the Data mode (USB-D or LSB-D) to transmit from an external program. In the other modes, the transmitter will key but the modulation will not be transmitted. The exceptions are DTR keyed modes. CW will work in the CW mode and FSK (not AFSK) RTTY will work in the RTTY mode.

*Note that audio is sent to the PC when it is in any mode, so you can use your digital mode PC software to see and decode PSK signals. But you **must** be in the USB-D DATA mode to transmit PSK from your digital mode PC software.*

➢ **WSJT-X for JT65 or FT8**

If you are using WSKT-X for JT65 or FT8 modes, the software 'Split operation' should be set to 'Rig.' See page 20. This allows the software to use 'Split' to change the transmit VFO frequency in order to reduce the possibility of audio harmonics affecting other FT8 users. For example, if you are transmitting at 600 Hz above the RF frequency of 14.074 MHz there could be harmonics at 1200 Hz and 1800 Hz. The function changes the VFO on transmit so that the audio tone is higher, making any possible audio harmonics fall outside of the transmitter's transmit bandwidth.

USING THE MEMORY CHANNELS

Using the memory channels is "a bit odd." For a start, there are two completely different ways to program or clear a memory slot and there are three different ways to select a memory channel.

The radio can store 99 frequencies and two scan edge memories P1 and P2. You can assign each memory slot to one of three scan groups. Each memory slot stores the mode, frequency, filter setting, CW break-in, VOX, compressor, and AGC settings.

You cannot enter a frequency directly into a memory slot. However, you can use an external memory manager to program memory slots via the CI-V interface.

➢ **Select a pre-programmed memory (method 1)**

You can select a memory slot using the up ▲ and down ▼ buttons to choose the memory slot you want to use, but it will not change the VFO frequency unless you use the V/M button or VFO/MEMORY sub-screen to select Memory mode.

Once you are in the memory mode the up ▲ and down ▼ buttons will change the VFO frequency and mode etc. If a memory name has been programmed it will be displayed above the 10s digit of the VFO frequency display. But only on the display with the large frequency digits.

Holding the up ▲ or down ▼ buttons will step more quickly through the 99 memory slots and two scan edges.

➢ **Select a pre-programmed memory (method 2)**

Touch and hold the MHz digits of the VFO display. Touch F-INP. Enter the memory number you want to select then touch MEMO. For example, <F-INP> <1> <2> <MEMO> will select memory slot 12.

If the radio is in VFO mode, this will have no effect other than setting the memory number on the right side of the display. Press the V/M button or VFO/MEMORY sub-screen to select Memory mode and then the VFO will change to the selected frequency, mode, and filter selection etc.

If the radio is already in MEMORY mode the VFO will change to the selected frequency, mode, and filter selection etc. immediately.

➢ **Select a pre-programmed memory (method 3)**

Press <MENU> <MEMORY>. You can use the up and down Soft Keys, turn the MULTI knob, or use the up ▲ or down ▼ buttons to display the 99 memories on pages 1 to 26.

Touch a memory entry to select a memory slot.

If the radio is in VFO mode, this will have no effect other than setting the memory number on the right side of the display. Press the V/M button or VFO/MEMORY sub-screen to select Memory mode and the VFO will change to the selected frequency, mode, and filter selection etc.

If the radio is in MEMORY mode the VFO will change to the selected frequency, mode, and filter selection etc.

➢ **Program or clear a memory slot (method 1)**

Set the radio to the frequency, mode and filter setting that you want to save in a memory slot.

Select a screen size that displays the large VFO numbers and the memory slot number displayed on the right side of the screen.

Use the up ▲ and down ▼ buttons to select the memory slot you want to use. If the memory position is empty it will have 'BLANK' under the number.

If the number has nothing underneath or *1, *2 or *3 then there is already something programmed into that slot.

Touch the memory slot number or VFO icon to open the VFO/MEMORY sub screen.

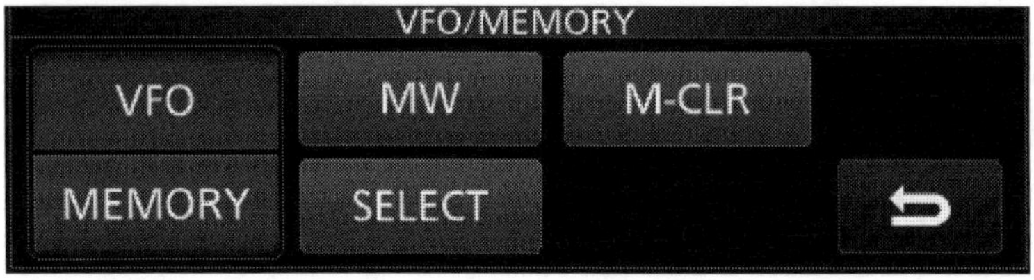

Touch VFO and then touch and hold MW to write the current VFO information into the selected memory slot. The word 'BLANK' under the memory slot number on the screen will disappear.

Touch MEMORY then touch (but don't touch and hold) SELECT to set a scan group number to the slot. Cycles through *1, *2, *3, and no scan group.

To clear a memory slot, touch MEMORY then touch and hold M-CLR. You don't have to clear the slot to overwrite it using the MW key.

Select VFO or MEMORY before leaving the screen with the return Soft Key or the EXIT button.

You can't use this screen to select a memory or set the contents to the VFO. "Weird Eh?" Icom could have put a 'touch and hold' function on the MEMORY Soft Key to display and select from the table of memory contents… but they didn't.

➢ **Program or clear a memory slot (method 2)**

Press <MENU> <MEMORY>. You can use the up and down Soft Keys, turn the MULTI knob, or use the up ▲ or down ▼ buttons to display the 99 memories on pages 1 to 26.

Touch and hold a memory entry or touch the 'file' icon on the right, to select a memory slot.

If the memory slot is blank there will only be one choice. You can save the current VFO settings into the memory slot, or not.

If the memory slot is already programmed, you get three choices. You can add or edit the memory name. You can overwrite the current information with the current VFO settings, or you can clear the slot. You don't have to clear the slot before you overwrite it. Exit the edit menu using the ⮌ Soft Key.

When the slot has been programmed you can touch the icon under the memory slot number to place the frequency into one of the three memory scan groups.

Touching the icon repeatedly cycles through *1, *2, *3, and no scan group. 'Touch and hold' clears the whole scan group or all three scan groups! Be careful.

Memory setting screen

In the image above memory slot, one has been set to 7.157 MHz LSB with a name of '40m Net.' The memory slot is in the second memory scan group.

Memory slot two has been set to 14.074 MHz USB data mode, with a name of '20m FT8,' The memory slot is in the first memory scan group.

➢ **Strange but true!**

If you touch and hold a line that has a frequency programmed into it, the first item on the popup menu is SELECT. You would think that would allow you to select the channel and pop it into the VFO… but it does not. It allows you to select the scan group, which you can also do by touching the star under the channel number.

Menu

MENU is the first of the buttons located below the touch screen. Initially, you will use the MENU button a lot as you set up the radio's many different options and features. Some of the icons that are displayed change according to the mode that the radio is in. For example, if the radio is set for SSB operation there is a VOICE option. In CW mode this is replaced by the KEYER option. In RTTY mode it becomes the DECODE option.

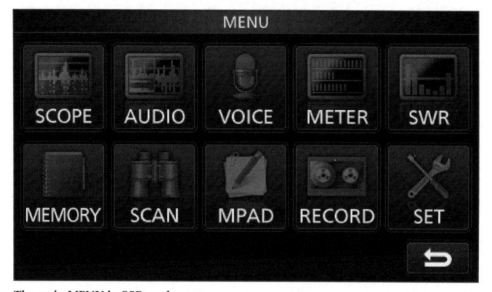

The main MENU in SSB mode.

FT8 PRESET

The February 2021 (V1.40) firmware update introduced a "one-touch," 'FT8 Preset' on a new second page of the main menu. It adds five one-touch "mode pre-sets." The idea was to quickly change the radio to the settings required for FT8, reflecting the popularity of the mode. The top pre-set is labelled 'Normal,' but I have renamed it to SSB because it does <u>not</u> return the radio to a previous setting. The second item is labelled 'FT8,' although the options it sets are suitable for most external digital modes. You can add three more pre-set arrangements of your own design.

Load the FT8 Preset before you start WSJT-X or another digital mode program. <MENU> <2> <PRESET> <FT8> <YES>

Unload it again when you have finished with the FT8 or other Preset. <MENU> <2> <PRESET> <UNLOAD> <YES>

Each PRESET can store, a Preset name, the mode, receiver filter, filter bandwidth, USB keying settings, CI-V settings, data mod type, data off mod type, TX bandwidth, speech compressor, and TX wide/mid/narrow.

You can select which items you want to store and unselect any irrelevant items. For example, I set up a Preset for FM receiving. It does not need any of the CI-V or transmitter settings.

There is more about what settings can be saved into a Preset back on page 27.

SCOPE

Touching the SCOPE Soft Key turns on the panadapter spectrum and waterfall display with the SCOPE Soft Key icons at the bottom. The size of the panadapter is determined by the EXPD Soft Key setting.

You can also activate the panadapter spectrum and waterfall display by press and holding the M.SCOPE button.

Pressing the M.SCOPE button (rather than press and holding it) cycles through two displays. One with a small panadapter with no Soft Key controls and the other with large VFO numbers and no panadapter. But for the following discussion, we need the display option that includes the Scope Soft Keys.

➢ **SCOPE Spectrum Scope Soft Keys**

<MENU> <SCOPE> or hold down the M.SCOPE button.

EDGE is displayed in the Fixed display mode it cycles through the four predefined band edges for the current frequency segment. SPAN is displayed in the Center display mode. It cycles through the eight span ranges.

	Function	Hold	Menu Setting / Notes
< 1 >	Change to Menu 2 selections.	None	
EDGE	In the FIX spectrum mode, EDGE cycles through four pre-set band edges per frequency zone.	None	To set the band edges <M.SCOPE> <hold EXPD/SET> <Fixed Edges>
SPAN	In Centre (CENT) spectrum mode, Span cycles through the Span settings.	Reset to ±2.5 kHz	±2.5 kHz, ±5 kHz, ±10 kHz, ±25 kHz, ±50 kHz, ±100 kHz, ±250 kHz, ±500 kHz
HOLD	Freezes the spectrum scope and waterfall.	None	
CENT/FIX	Toggles the FIX and CENT span modes.	Scroll Modes	FIX mode displays the spectrum between pre-defined band edges. CENT displays the spectrum below and above the active VFO frequency. SCROLL-F displays the spectrum with a span defined by the pre-defined band edges. SCROLL-C displays the spectrum with a span defined by the SCAN bandwidth.
EXPD/SET	Touch EXPD/SET to turn on or off the expanded spectrum scope display. Touch and hold EXPD/SET to enter the screen setup menu.	Enter SET menu	See next table for the SET menu

14.080	14.081	14.082	14.083	14.084
< 2 >	REF	SPEED	MARKER	EXPD/SET

< 2 >	Change to Menu 1 selections.	None	
REF	Allows adjustment of the Spectrum display reference level.	None	Use the Main VFO knob to adjust REF level. Touch REF again to return to Menu 2. Touch and hold DEF to set REF to 0 dB.
SPEED	Cycle through FAST, MID, SLOW panadapter and waterfall speeds.	None	
MARKER	Displays the orange (T) transmit frequency marker. In the FIX spectrum mode, the green (R) receive frequency marker is always displayed.		>> at the right side of the panadapter or << at the left side of the panadapter indicates that the marker is lower or higher than the panadapter can show on its current setting.

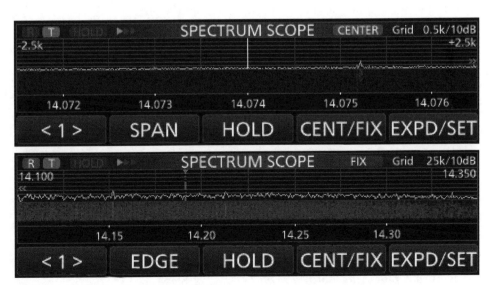

The Centre and Fixed panadapter modes

➢ **SCOPE Spectrum Scope EXPD/SET settings**

Select <MENU> <SCOPE> or hold the M.SCOPE button for one second, until the large or medium panadapter is displayed. Then hold the EXPD/SET Soft Key down for one second to display the 16 scope settings. See the notes about the carrier point display after the table. I believe that these settings are important.

Setting	IC-7300 default	ZL3DW setting	My Setting
Scope during TX (CENTER) Sets the scope to display the transmitted signal while you are transmitting.	On	On	
Max Hold Sets the peak hold function on the spectrum display	10-second hold	Off	
CENTRE Type Display Selects the relationship between the CENT scope line and the receiver filter	Filter Center 1	Carrier Point Center (Abs Freq) [2 & 3]	
Marker Position (FIX Type/SCROLL Type) Selects the relationship between the FIX scope marker line and the receiver filter	Carrier Point 2	Carrier Point 2	
VBW (video bandwidth)	Narrow	Narrow	
Averaging smooths the spectrum display by averaging 2, 3, or 4 sweeps.	OFF	Averaged over 3 sweeps	
Waveform type. Filled or filled with a line. See 'Useful Tips' to emulate a line display.	Fill	Fill+Line (with colour modification)	
Waveform Color (Current)	R:172 G:191 B:191	R:0 G:0 B:57	
Waveform Color (Line)	R:56 G:24 B:0	R:200 G:200 B:200	
Waveform Color (Max Hold)	R:45 G:86 B:115	R:45 G:86 B:115	

Waterfall display	ON	ON	
Waterfall speed	MID	MID	
Waterfall Size (Expand Screen)	MID	SMALL	
Waterfall Peak Color Level	Grid 8	Grid 8	
Waterfall Marker Auto-hide	ON	OFF	
Fixed Edges	Displays 13 frequency zones. Each zone has four band edge settings. See 'Setting up the panadapter band edges' on page 65.		

1. Filter centre places the VFO frequency marker in the middle of the filter bandwidth, at the centre of a USB or LSB signal. In my opinion, this is just plain wrong.

2. Carrier Point (Center) places the VFO frequency marker at the carrier point. The left side of an upper sideband (USB) signal or the right side of a lower sideband (LSB) signal. In FIX mode bottom of the display shows the actual frequency. In CENT mode, the bottom of the display shows the span in kHz.

3. Carrier Point Center (Abs Freq) places the VFO frequency marker at the carrier point. The left side of an upper sideband (USB) signal or the right side of a lower sideband (LSB) signal. The bottom of the display shows the actual frequency in MHz.

AUDIO

➢ **AUDIO sub-screen Scope Soft Keys**

Selecting <MENU> <AUDIO> displays the received audio level and spectrum on the AUDIO SCOPE. While transmitting it displays the transmit audio level and spectrum. The audio scope will not display received audio if the receiver is muted. You can see the effect of the filter passband on the width of the audio spectrum display and the effect of the NR noise reduction filter and the manual notch filter.

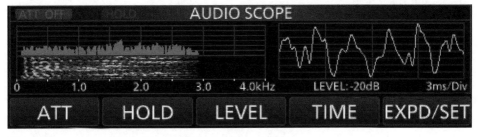

The audio scope works when receiving or transmitting

Soft Key	Function	Hold	Menu Setting / Notes
ATT	Adds attenuation to reduce the level of the audio spectrum and the brightness of the waterfall.	Resets to 0 dB.	Selectable 0dB, 10 dB, 20 dB, or 30 dB.
HOLD	Freezes the audio spectrum scope, the waterfall, and the oscilloscope.	None	
LEVEL	Changes the gain of the oscilloscope display.	None	Selectable 0dB, -10 dB, -20 dB, or -30 dB.
TIME	Changes the time-base of the oscilloscope display.	None	Selectable 1 ms, 3 ms, 10 ms, 30 ms, 100 ms, or 300 ms per division.
EXPD/SET	EXPD makes the audio spectrum and oscilloscope display taller.	SET mode see table below	See the Audio Scope SET mode table below

> ➤ **AUDIO Scope SET settings**

Select <MENU> <AUDIO> <SET> to display the Audio settings. You can set the spectrum, waterfall and oscilloscope trace colour, or turn the waterfall off.

Setting	IC-7300 default	ZL3DW setting	My Setting
FFT Scope Waveform Type	Fill	Line	
FFT Scope Waveform Color	R:51 G:153 B:255	R:51 G:153 B:255	
FFT Scope Waterfall Display	ON	ON	
Oscilloscope Waveform Color	R:0 G:255 B:0	R:0 G:255 B:0	

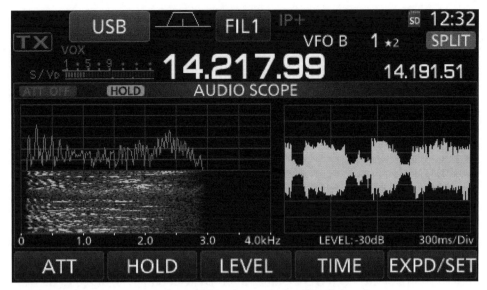

The Audio Scope expanded display showing a received SSB signal

VOICE

➤ **VOICE sub-screen Soft Keys**

The Voice option is only available when the active VFO is in a voice mode or a voice-data mode. Touch T1 to T8 to send the voice messages. You must have an SD card installed to use the voice keyer. Each of the eight 'voice messages' can be up to one and a half minutes long. Mine are all less than ten seconds long.

To record a voice message, select <MENU> <VOICE> <REC/SET> <REC>.

- Touch the message number T1 – T8 that you want to record. A recording screen will pop-up. The three buttons are Record, Playback, and Stop.

- You can record the message using your normal microphone. Do not press the PTT button. You don't need to transmit to make a recording. This is handy. I recorded a new message during a contest the other day.

  o You set the recording level by speaking into the microphone in the same way that you would "on the air." Touch the MIC GAIN control and set the level so that the peak level hold line just hits 100%. Don't worry if the occasional peak is higher than 100%. The transmitter limiter will deal with that.

  o When you have set the Mic Level you can proceed with recording the messages. I write down exactly what I want to record. It really helps. Practice it a couple of times.

- o Touch the red 'record' icon and speak normally into the microphone. Touch the square 'stop' icon immediately after you stop speaking. It's best if you have your finger ready, poised over the button.

- o You can play back the recording by touching the triangle 'play' button.

- o If you don't like the recording, just record a new message. It will overwrite the old message automatically.

- o When you are happy with the recording. Touch the return ↰ icon and record the next message in the same way.

- You have to make the recording before adding a name to the T button. To add a name to the recording, touch and hold the message number (T1 – T8). Then select <Edit Name>. Type the name of the message, for Example 'CQ short,' or 'my QTH.' The names are displayed on the eight 'voice message buttons.' They can be up to 16 characters long, but the button can only display about eight characters.

- Make sure to touch <ENT> or your changes will be lost when you exit the screen. To exit the recording screen, touch the return ↰ icon or press the EXIT button.

Soft Key	Function	Hold	Menu Setting / Notes
T1 to T8	T1 to T8 sends the voice message.	Repeat	
REC	Record voice messages (see above)	None	<REC/SET> <REC>
Auto Monitor	'ON' lets you hear the voice message on playback	None	<REC/SET> <SET> <Auto Monitor>
Repeat Time	Set the time between message repeats. The default is 5 seconds.	None	<REC/SET> <SET> <Repeat Time>
TX LEVEL	Adjust TX LEVEL while the message is playing to get the same RF power as you would if you were speaking into the microphone. i.e. peaking to 100 Watts.	None	Use the main VFO knob to adjust. The default level is 50%.

KEYER

The CW Keyer menu is only visible if the radio is in CW mode. Select <MENU> <KEYER>. It is used to set all CW functions and to send the eight CW messages.

➤ **KEYER sub-screen Soft Keys**

Touching one of the **M1 to M8 Soft Keys** sends the CW message.

Touch and hold one of the **M1 to M8 Soft Keys** to keep sending the message until you stop it by touching the Soft Key again or by sending with the key or paddle.

You can also use the F1 to F4 keys on an external keyboard or buttons connected via the microphone connector. Set <MENU> <SET> <Connectors> <External Keypad> <KEYER> <ON>.

Although there is a counter to increment the contest report number, there is no way to enter a callsign into a QSO or contest exchange other than editing the message. For contests, it's probably easier to just use N1MM Logger +.

The -1 Soft Key: If you get a busted contest QSO you can decrement the counter by touching -1 on the Keyer message screen. The next M2 message will repeat the previously sent number. You can't manually increment the counter from this screen, but you can set it to any number you like using <MENU> <KEYER> <EDIT/SET> <001 SET> <Present Number>.

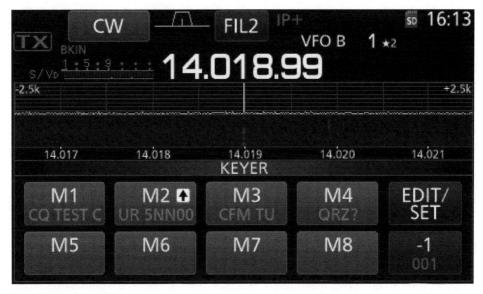

The CW message keyer screen

The up arrow on M2 indicates that it will send the contest number.

To **edit a CW message**, select <MENU> <KEYER> <EDIT/SET> <EDIT>.

- Touch the message that you want to edit and then touch the <EDIT> icon. A keyboard will appear on the touch screen.

- CW messages can be up to 70 characters long.

- A Caret ^ symbol removes the space between two letters, for example, ^AR.

- Press ENT to save your changes.

- Message M2 has an upward pointing arrow. It signifies that this message slot has an auto-incrementing counter used for contest exchanges.

  In operation, the * is replaced with the contest number. The default message for the M2 message is UR 5NN * BK, which will send UR 5NN 001 BK, etc.

  Only one message can send the contest number. You can move the contest number (*) to another message but you must delete the * in the M2 message first. Also, if you do change the message that has the contest number (*), you have to tell the radio which message now has the * by changing the trigger. <EDIT/SET> <001 SET> <Count Up Trigger>.

  If you don't work contests you can delete the * and use the M2 message slot for a different message.

- When you have finished editing all the messages touch the return ↺ icon to exit.

001 SET: On the <MENU> <KEYER> <EDIT/SET> <001 SET> menu you can set

a) The **number style** used for automatic contest numbers.

  a. Normal numbers, 001 etc.

  b. ANO style A=1, N=9, O=0

  c. ANT style A=1, N=9, T=0

  d. NO style N=9, O=0

  e. NT style N=9, T=0

b) **Count Up Trigger.** Sets which macro is using the automatic 'Count Up' feature. The default is M2. It must be the macro that has the contest number star (*) in the text.

c) The **Present Number**, usually 001. Note that if you get a busted contest QSO you can decrement the counter by touching -1 on the Keyer message screen.

CW-KEY SET: On the <MENU> <KEYER> <EDIT/SET> <CW-KEY SET> menu you can set the following:

Setting	IC-7300 default	ZL3DW setting	My Setting
Sidetone Level Turn it all the way down if you don't want sidetone.	50%	50%	
Sidetone Level Limit This apparently disables the sidetone if you turn the volume control up past the sidetone level.	ON	ON	
Keyer Repeat Time Sets the time before a memory message is re-sent if you have selected the automatic option. A red 'wait' icon is displayed between transmissions.	2 seconds	2 seconds	
Dot/Dash ratio The default is a 'Dah' that is three times the length of a 'Dot' but this can be adjusted between 2.8 and 4.5.	1:1.3	1:1.3	
Rise Time Sets the CW rise time. Select 2 ms, 4 ms, or 8 ms. The default is 4 ms.	4 ms	4 ms	
Paddle Polarity can be set to normal or reverse to suit your operating style.	Normal	Normal	
Key Type can be set to Paddle, Bug, or Straight Key.	Paddle	Paddle	
MIC Up/Down Keyer allows the up-down buttons on the hand microphone to send CW. Up (left) is Dits and Down (right) is Dahs.	OFF	OFF	

DECODE

The DECODE icon is visible when the radio is in RTTY mode. It activates the internal RTTY decoder and message memory keys.

➢ **DECODE sub-screen and Soft Keys**

To edit an RTTY message, select RTTY then select <MENU> <DECODE> <TX MEM> <EDIT>.

- Touch the message that you want to edit and then touch the <EDIT> icon. A keyboard will appear on the touch screen.

- Messages can be up to 70 characters.

- Inserting a carriage return character ↵ causes the following text to appear on a new line at the receiving station.

- Press ENT to save your changes.

Soft Key	Function	Hold	Menu Setting / Notes
< 1 >	Change to Menu 2 selections.	None	
HOLD	Touch HOLD to halt the decoder. This stops text scrolling past so you have time to read a callsign etc.	None	
CLR	Touch and hold to clear the received text area.	Clear	Touch and hold to clear the received text area.
TX MEM	Opens the eight RTTY message memories.	None	They can be edited by touching the EDIT icon.
< 2 >	Change to Menu 1 selections.	None	
LOG	Opens a sub-menu. You can turn on the logging function and change it between a text file and an HTML file.	None	The text log records all decoded text and the text that you send. See the LOG table below.

LOG VIEW	Opens a file dialogue where you can select an RTTY text log to view.	None	The text log records all decoded text and the text that you send. Useful data for your station log entries.
ADJ	Touching ADJ allows you to use the Main VFO knob to adjust the decode threshold. Touch ADJ again to exit.	None	The default threshold is 8. Even a setting of 15 does not stop false decodes.
DEF	Touch and hold DEF to return to the default decoder threshold of 8. Touch ADJ to exit.	Function	DEF is only visible on the ADJ sub-menu.
EXPD/SET	Increases the size of the Decoder screen so that the send and received text areas are larger. This overlays the Spectrum Scope. But touching EXPAND again restores it.	SET	SET opens a menu with many RTTY format, FFT display, and decoder options.

➤ **The RTTY decode LOG submenu.**

Setting	IC-7300 default	ZL3DW setting	My Setting
Decode Log When the Log is on, all decoded characters are recorded until you go back to this menu and stop it. A red dot in the top left of the RTTY Decode screen, next to the TX icon, indicates that the log is recording.	OFF	OFF	
File type Text or HTML. HTML is better because it is in colour.	Text	HTML	

Setting			
Time Stamp Adds a time stamp date and time at the start of each receive and transmit period. The date format is YYYYMMDD	ON	ON	
Time Stamp (Time) Sets the timestamp to local time or UTC. Note that this will be affected if you have set the radio clock to UTC. In my opinion, all logs should be in UTC, but it's up to you.	Local	UTC	
Time Stamp (Frequency) Adds the operating frequency to the log. "Yeah, why not?"	ON	ON	

➤ **The RTTY decode SET submenu.**

Touch and hold any menu item to reset the item to the Icom default.

Setting	IC-7300 default	ZL3DW setting	My Setting
FFT Scope Averaging Adding averaging makes the tuning spectrum scope smoother, but the delay that is introduced makes tuning in RTTY signals more difficult. Can average 2, 3, or 4 sweeps.	OFF	OFF	
FFT Waveform Color The default is pale blue, but you can adjust it if you want to. Some folks like a green or a grey scale image.	R:51, G:153, B255	R:51, G:153, B255	

Decode USOS	ON	ON	
USOS stands for unshift on space. Standard Baudot code will stay on characters or letters until a shift character is received. This function assumes that a letter is more likely to follow a space than a number is, and automatically shifts the receiver back to letters. Don't change the setting unless you are having decoding problems or are receiving secret messages sent in 5 number groups.			
Decode New Line Code	CR, LF, or	CR, LF, or	
This changes the decode screen. Setting a new line only when a carriage and line feed is received (CR+LF), or when a CR, LF, or a CR+LF character is received. The default option usually makes the received text easier to receive. But it can break up sentences.	CR+LF	CR+LF	

METER

Selecting <MENU> <METER> displays the Multi-Function meter. You can also enable the meter by touch and holding the meter display on the touch screen.

The Multi-Function meter works best if you use the M.SCOPE button to select the screen with no panadapter display. That way you get the large S meter/RF Power meter.

When receiving, the Multi-Function meter shows you the S meter, supply voltage, and temperature of the final MOS-FET stage. When the radio is transmitting the meter shows you all the transmit metering at the same time. This is very handy when you are setting the Mic Gain and Compression controls.

- Po (RF power output 0 – 100%)
- ALC (voice peaks should always be within the red zone)

- COMP (with the compressor turned on voice peaks should always be within between 5 and 20 dB – never more than 20 dB)

- SWR (Standing wave ratio. Hopefully, less than 1.5)

- I_D (current being drawn by the final amplifier MOS-FETs)

- V_D (drain voltage on the final amplifier MOS-FETs)

- Temperature (measured at the final amplifier MOS-FETs).

<MENU> <SET> <Display> <Meter Peak Hold> adds a peak hold function to the S meter while receiving and the Po output power meter while transmitting. The default is ON.

SWR

The SWR plot function is a rather unique feature of the IC-7300. It allows you to carry out a check of your antenna system. While it is not as effective as using a dedicated antenna analyser it will give you an idea of the current state of your antenna. The SWR mode lets you plot the antenna SWR across a nominated frequency span.

By the way, you can check the SWR at your specific transmit frequency any time you are transmitting by selecting the SWR panel meter, or the multi-function meter. Touch the meter scale to cycle through the various mode until it says SWR at the left of the lower meter scale. It's best if you use one of the screens that show the large meter scale. Touch and hold the meter scale to show the multi-function meter which has a separate SWR meter.

➢ **Making an SWR plot**

1. Set the transmit power to 30% to avoid the possibility of damaging the transceiver through high SWR. <MULTI> <RF POWER> = 30%.

2. If you want a real plot of your antenna system, turn the IC-7300 antenna tuner off, and set any external tuner to bypass. If you don't, the tuner may tune to the first or any of the transmissions causing misleading results.

3. Set the main VFO to the centre of the frequency band that you want to measure. For example, 14.200 MHz on the 20m band.

4. The radio will not transmit on frequencies outside of the amateur bands. Note that the VFO will not tune after the recording Soft Key ▷ ☐ has been selected. Touch it again to regain VFO control.

5. Select a STEP size, probably 10 kHz or 50 kHz unless you have a wideband antenna, or you are testing on the 10m or 6m band. 50 kHz with 7 bars will scan from 14.050 to 14.350 MHz.

6. Select the number of bars to display. They will be spaced at the STEP frequency. For example, a setting of three bars and a step size of 50 kHz, makes measurements at 14.150, 14.200, and 14.250 MHz. A setting of five bars and a step size of 50 kHz, makes measurements at 14.100, 14.150, 14.200, 14.250 and 14.300 MHz.

7. Touch the ▷ ☐ Soft Key so that it becomes ▷ ◼. Press the microphone PTT but don't talk into the mic. The transmitter will transmit. Release the mic after a second or two. About the time it takes for the fan to wind up. You will see a small yellow triangle step from the first bar to the next frequency position. You might not see any indication on the SWR scale. Or you might see a blue bar or a blue and red bar.

8. Press the mic PTT again, and release after a second or two. Repeat until the plot is finished and the small yellow triangle returns to the centre (VFO) frequency position.

9. You should see a U shape of bars, or better still no bars. See the images below. SWR below 1.5 is indicated with no bar or a blue bar. SWR higher than 1.5 is indicated with a blue and red bar.

10. If you want to use the antenna at frequencies where the SWR is over 1.5 you will have to use an antenna tuner, either the one in the radio or an external one. The internal tuner will match SWR of 3:1 or less. External tuners can usually match a wider range. Alternatively, you could construct or trim your antenna so that it matches the transceiver at the wanted frequency of operation.

11. When you are sure that you have finished. Press the EXIT button to exit the SWR screen and set the transmit power back to 100%. <MULTI> <RF POWER> = 100%.

12. Pressing <MENU> <SWR> returns you to the last measurement that you made. Touch and hold the RECALL Soft Key to set the VFO to the SWR plot centre frequency.

SWR plot. 13 bars with 10 kHz steps

SWR plot. 7 bars with 50 kHz steps

MEMORY

The MEMORY Soft Key opens the main memory screen. You can save the current VFO frequency including the current mode and filter, delete a stored memory, or edit the name of any existing saved memory. Rather strangely unless you happen to be in Memory mode rather than VFO mode, touching a memory returns you to the main operating screen but does not change the VFO to the selected memory frequency. The radio can store 99 frequencies including the mode and filter that was in use, plus two scan edge memories P1 and P2.

There is also the facility to make any saved memory a member of one of the three scan groups. The memory scan mode can scan the memories in a nominated scan group or scan all of the memory slots.

➢ **MEMORY sub-screen Soft Keys**

You can store the current VFO frequency to a memory slot by touch and holding any of the memory slots. If you select an already occupied slot you have the choice of overwriting the contents with the current VFO settings.

Memory setting screen

Each memory entry looks like the image above.

- You can use the up and down Soft Keys, turn the MULTI knob, or use the up ▲ or down ▼ buttons to display the 99 memories on pages 1 to 26.
- The number on the left is the memory slot number from 1-99.

- The *3 is a scan group number. There are three scan groups. You can scan through any of the scan groups, all three scan groups, or all 99 memory slots. Touch the leftmost entry box to cycle through; *1, *2, *3, or none.

- Touch and hold a memory entry or touch the 'file' icon on the right, to select a memory slot.

- If the memory slot is blank there will only be one choice. You can save the current VFO settings into the memory slot, or not.

- If the memory slot is already programmed, you get three choices. You can add or edit the memory name. You can overwrite the current information with the current VFO settings, or you can clear the slot. You don't have to clear the slot before you overwrite it. Exit the edit menu using the ↺ Soft Key.

- When the slot has been programmed you can touch the icon under the memory slot number to place the frequency into one of the three memory scan groups. Cycles through *1, *2, *3, and no scan group. Touch and hold clears the whole scan group or all three scan groups! Be careful.

➤ **The alternative memory management system**

You can also save memory contents using the memory submenu accessed by touching the memory number on the touch screen. I am puzzled as to why the radio has two completely different methods of managing the memories. For more on the alternative memory management system see; 'Using the memory channels' page 84.

The quick way to save a frequency into a memory slot

> **FM Repeater Channels**

There is no specific memory storage for FM channels. But you can use the standard memory channels and name the memory channel with the repeater identification. The standard memory channels will store the repeater split offset and CTCSS tone so set them before saving the channel to a memory slot.

> **External memory managers**

The WCS-7610 Programming Software by RT Systems can read memories and menu settings from the radio. You can edit them in the PC software and then upload them back to the radio over the USB cable. The program uses the CI-V command interface. Please note that I have not tried this software. I am not endorsing it, just letting you know that it exists.

SCAN

Selecting <MENU> <SCAN> displays the SCAN Menu. If you open the scan screen when the panadapter is displayed, it will shrink to allow for the scan sub-menu screen. If the panadapter is not displayed, you can press M.SCOPE to display it.

If you have the spectrum display set to FIX you can see the green marker moving across the band as the VFO scans. Touching the screen anywhere, keying PTT from the Mic, or touching the scan button again will halt the scan.

Set the squelch so that the radio is squelched before using the Scan mode because the scan will stop when a signal is encountered, and the squelch opens. If Scan Resume is set to 'on' the scan will resume after a time.

Selecting <MENU> <SCAN> displays the SCAN Menu. Touching the screen, PTT from the Mic, or touching the scan button gain will halt a scan that is in progress.

The radio has three scan modes. You can scan between the P1 and P2 memory frequencies, you can scan through programmed memories, or you can scan a range around the current VFO frequency.

- The ◢F Soft Key starts a scan from just below the current VFO frequency to just above it. You can set the scan range from ±5 kHz to ±1 MHz using the ◢F SPAN Soft Key.

- Touch PROG to scan between the P1 and P2 frequencies. PROGRAM SCAN flashes while the scan is progressing and the decimal points in the frequency display flash.

- MEMO replaces PROG if the radio is in Memory mode instead of VFO mode. It is used to scan through selected memories from the 99 stored memories.

Set the squelch so that the radio is squelched before using the Scan mode and the Scan will stop when a signal is encountered. You can start the scan again by touching the PROG or ◢F Soft Key. Alternatively, if Scan Resume is set to 'ON' the scan will resume after a time. If you touch the FINE Soft Key **after** a scan has started, the scan will slow down but not stop when the squelch opens. After it passes through the signal and the squelch closes again, the scan speeds up again. I like the ◢F scan mode with 'Scan Resume' set to ON.

In effect, there are four scan speeds. The scan uses the VFO tuning step and the SET menu lets you select either a Fast or Slow scan speed. The fastest scan is when the VFO is in the 1 kHz step mode and the scan speed is set to fast. The slowest scan speed is when the VFO is set to 1 Hz steps and scan speed is set to slow.

Scan controls VFO mode

Scan controls in memory mode while scanning

Soft Key	Function	Hold	Menu Setting / Notes
PROG	Touch PROG to scan between the P1 and P2 frequencies. PROGRAM SCAN flashes while the scan is progressing and the decimal points in the frequency display flash.	None	P1 and P2 can be set in the memory screen like any other memory slot. <MENU> <MEMORY>

◢F	Span frequencies below and above the current VFO setting. ◢F SCAN flashes while the scan is progressing and the decimal points in the frequency display flash.	None	The range of the span is set by touching the ◢F SPAN control.
FINE	Touch FINE while a scan is in progress and the scan rate will slow but not pause when a signal opens the receiver squelch.		
◢F SPAN	Sets the range of the ◢F scan.	None	±5 kHz, ±10 kHz, ±20 kHz, ±50 kHz, ±100 kHz, ±500 kHz, ±1 MHz.
RECALL	Touch and hold to reset the VFO to the original frequency.	VFO recall function	Touching the screen, PTT from the Mic, or touching the scan button gain will halt the scan.
SET	Sub-menu to change Scan speed and Scan resume function	None	FAST/SLOW and ON/OFF
MEMO	MEMO replaces PROG if the radio is in Memory mode instead of VFO mode. It is used to scan through selected memories from the 99 stored memories.	None	Memo steps through all the stored memories. Or through the stored memories that are tagged with any of the three scan group tags, *1, *2, *3.
SELECT	When not scanning, SELECT changes the scan group of the current memory channel. While scanning SELECT selects all channel or scan group only scan. This is buggy	Clears one or all scan groups	Touch and hold clears one or all scan group settings. Be careful.

	and sometimes it stops the scan instead. Using SEL no. then SELECT seems to fix the issue.		
SEL No.	Selects the group of memory channels that will be scanned. *1, *2, or *3, or *1, *2, and *3.	None	SEL No. is only displayed in memory scan mode while the scan is running.

MPAD

Selecting <MENU> <MPAD> displays the MEMO PAD menu. The memory pad is used for short-term storage of a frequency and mode that you might want to return to later. As new entries are added to the top of the stack, the oldest disappear. The memory pad can store either 5 or 10 frequencies, depending on the setting, <MENU> <SET> <Function> <Memo Pad Quantity>.

I am not sure why Icom bothers to include this sub-screen. It gives you an opportunity to see what is saved in each memory pad slot, but the next time you press and hold the MPAD button all the entries will move down one slot. Pressing the MPAD button has the same effect as the up and down keys, cycling through each stored frequency. The screen does give you the option of deleting one or all the stored memories, but why bother?

Soft Key	Function	Hold	Menu Setting / Notes
▲	Step up through the stack and the current VFO frequency.	None	The MPAD stack holds 5 (or 10) channels. Up/Down cycles through them plus the VFO frequency.
▼	Step down through the stack and the current VFO frequency.	None	
DEL	Touch and hold to delete the currently highlighted entry.	Delete entry	
DEL ALL	Touch and hold to delete all of the memory pad entries.	Delete all entries	

| EXPAND | Increases the size of the MEMO PAD so that you can see all 10 entries. | None | This overlays the Spectrum Scope. But touching EXPAND again restores it. |

RECORD

The QSO recorder records the receiver. You can start recording by selecting <<REC Start>> on this screen, or via the QUICK button.

A red dot ● REC icon at the top of the screen to the left of the clock indicates that the recorder is recording. Also, the blue SD card icon will flash indicating a 'write' to the SD card. If the squelch is closed the recorder will pause, replacing the red dot with a pause symbol, ‖ REC icon.

Touch the icon to stop recording. Or you can go back into the <MENU> <RECORD> or <QUICK> menus and select <<REC Stop>>.

Soft Key	Function	Hold	Menu Setting / Notes
<<REC Start>>	Touch <<REC Start>> to start recording audio to a file on the SD card. A menu choice allows you to record just the received signal or both the received and your transmit signal.	None	● REC means recording ‖ REC means paused (squelch closed)
<<REC Stop>>	Touch <<REC Stop>> to stop recording	None	
Play Files	Allows you to select a recorded file and play it back.	None	No, you cannot use the recording to modulate the radio. Touch and hold a folder or a file to delete it.
Recorder Set	See the next table.	None	
Player Set	Sets how far the audio will skip ahead if you touch the fast forward icon during playback. You can skip back too.	None	3, 5, 10, or 30 seconds. Touch and hold sets the time back to default 10 seconds.

➤ **Recorder Set – sub-menu**

Setting	IC-7300 default	ZL3DW setting	My Setting
REC Mode. Record the received audio or both the received and transmitted audio.	TX&RX	TX&RX	
TX REC Audio. Record microphone audio (Direct) or the Transmit Monitor audio (Monitor).	Direct	Direct	
RX REC Condition. Record only when the receiver squelch is open or record all the time.	Squelch Auto	Squelch Auto	
File Split. Create a new file when you transmit, or the squelch opens (ON). Or keep recording until manually stopped or the 2Gb maximum file size is reached (OFF).	ON	ON	
PTT Auto REC. If set to 'ON' a recording will start every time you transmit. If 'OFF,' recording will only start if you start it.	OFF	OFF	
PRE-REC for PTT Auto REC. If 'automatic recording' is ON, Pre-Rec will record up to 15 seconds of the received audio before you press the PTT on the microphone.	10 seconds	10 seconds	

SET MENUS

SET is used to access all of the deep menu settings. I have covered the important ones in 'Setting up the radio' and other chapters. The following is a quick summary.

Tone Control/TBW

- RX. Set audio high pass and low pass filters for the transmit modes. Or set the bass and treble for the AM, FM, and SSB modes. See page 13

- TX. Set the bass and treble for AM, FM, and SSB. Set transmit bandwidth for the SSB and SSB Data modes. See page 11.

Function

- Beep settings (four settings)
- Band Edges (User Band Edges). Set three band edges for each of the thirteen frequency ranges. Hidden function, see page 43
- RF/SQL Control. RF and Squelch, Squelch only, Auto
- MF Band ATT. Attenuator for the medium wave 'AM' broadcast band
- TX Delay. TX delay from SEND PTT to transmit. For HF and 6m bands. Set to at least 10ms if using a linear amplifier. Longer if it is relay switched
- CI-V timeout. Default 30 minutes
- Split settings
- Antenna Tuner. Tune on PTT start, the function of the TUNER button, or reset the tuner to default settings, losing saved data
- RTTY settings
- SPEECH button settings
- Dial LOCK button settings
- Memo Pad Quantity, 5 or 10 memory slots
- Main Dial Auto TS. Increases tuning step if you move the VFO knob quickly
- Mic Up/Down scan speed
- Quick RIT / XIT Clear (default OFF)
- Notch Filter auto/manual SSB mode
- Notch Filter auto/manual AM mode
- SSB/CW synchronous tuning. Sets the VFO to the CW offset when receiving CW on SSB mode. The default is OFF
- CW 'normal side' LSB or USB (default LSB)
- Screen capture settings
- Onscreen keyboard type (QWERTY or 10 key phone pad)
- Full keyboard layout (English, German, or French). This may be different on your radio due to market zones
- Calibration marker (OFF)
- Reference frequency adjustment. Don't touch this unless you know what you are doing. It is a factory pre-set. See Icom Full manual page 13-4

Connectors

- ACC and USB output AF (audio) or IF (12 kHz) output

- ACC and USB AF output level

- ACC and USB AF squelch. The default is 'OFF' so the audio to the computer is always available. The 'ON' mode only sends audio to the computer if the squelch is open

- ACC and USB AF Beep/Speech. OFF: beeps and SPEECH button audio are not sent to the computer. ON: beeps and SPEECH button audio are sent to the computer

- ACC and USB IF output level

- ACC modulation level. Audio input from peripheral on ACC connector

- USB modulation level. Audio input from PC on the USB cable

- DATA OFF MOD. Sets the source of modulation audio when the radio is not in a data mode. The default is MIC and ACC

- DATA MOD. Sets the source of modulation audio when the radio is in a data mode. The default is the USB cable

- External keypad enables the sending of message memories using buttons or a keypad connected to the microphone connector

- CI-V. See the CI-V settings table on page 118

- USB serial function. Use the USB cable for CI-V control or for decoded RTTY data

- RTTY decode Baud rate. Sets the data rate for RTTY to the PC

- USB Send Keying sets the COM port RTS and DTR settings

Display

- LCD Backlight sets the display brightness

- Display Type sets a black (default) background or s blue background

- Display Font sets either 'basic' (default) or 'round' an Arial type font

- Meter Peak Hold: What it says on the box, (ON)

- Memory name: Memory names are only displayed on the non-panadapter display, (ON)

- MN-Q Popup: (very odd) sets whether the radio will show the manual notch width popup when you select manual notch

- BW Popup (PBT) sets whether the radio will show the PBT popup when you change the PBT controls

- BW Popup (FIL) sets whether the radio will show Filter popup when you change the touch the filter Soft Key

- Screen Saver timeout

- Icom opening message and splash screen settings

- Display language: English or Japanese

Time Set

- Date and time setting

- UTC offset

SD Card

- Display a stored screen capture on the display

- Firmware update from the SD card

- Format the SD card

- Unmount the SD card

Etc. Others

- Information: Firmware versions

- Touch screen calibration

- Radio Reset – 'Partial Reset' sets operating settings back to factory default. It does not erase memory channels, callsign, keyer memories, band edges, or ref adjust

- Radio Reset – 'All Reset' sets everything back to factory defaults

- Emergency - enables you to use the antenna tuner when the antenna system has an SWR higher than 3:1 provided the output power is reduced to 50 watts or less.

SPECIAL SET MENU ITEMS

This section covers the items buried in the menu structure that you will probably need to adjust. These include

- The USB 'send' and keying settings (RTS and DTR) control lines for PTT and CW

- The CI-V settings (data rate and Icom address). There is no COM port allocation as these are set by the Icom driver software

- Audio level settings (for the USB cable)

➢ **USB SEND/Keying settings**

Select <MENU> <SET> <Connectors> <<USB SEND/Keying> to display the settings.

The USB 'send and keying' settings configure the way that PC digital mode software sends CW and PTT signals to the radio. You can use either line for the transmit PTT as long as you use the other line for CW.

I use RTS (ready to send) for the SEND (PTT) control and DTR (device terminal ready) for the CW. It is important that your digital mode software uses the same control lines. The USB Keying (RTTY) setting is for FSK RTTY operation (as opposed to AFSK RTTY). It should be set the same as CW, on my radio that is DTR.

For more information, see 'Radio and COM Port device setting' on page 18 and USB Send/keying settings' on page 22.

Setting	IC-7300 default	ZL3DW setting	My Setting
USB SEND	Off	RTS	
USB Keying (CW)	Off	DTR	
USB Keying (RTTY)	Off	DTR	
Inhibit timer at USB connection	ON	ON	

The 'Inhibit timer at USB connection' function is used to stop the radio sending a SEND, transmit keying signal when the USB cable is first plugged in. The manual states that this is only a problem if you are using old firmware. But I left it set to the default ON setting.

➢ **CI-V Settings**

<MENU> <SET> <Connectors> <CI-V>

These settings are for the CI-V (CAT) connection between the radio and digital mode software running on your PC. I experimented and found that the default settings are best.

	Setting	IC-7300 default	ZL3DW setting	My Setting
1	CI-V Baud Rate	Auto	Auto	
2	CI-V Address	94h	94h	
3	CI-V Transceive	ON	ON	
4	CI-V USB→REMOTE	00h	00h	
5	CI-V Output (for ANT)	OFF	OFF	
6	CI-V USB Port	Unlink from REMOTE	Unlink from REMOTE	
7	CI-V USB Baud Rate	Auto	Auto	
8	CI-V USB Echo Back	OFF	OFF	

1. CI-V Baud rate sets the data rate to a device connected to the REMOTE jack. If the 'Link to [REMOTE]' option is chosen, the CI-V Baud rate also sets the data rate to a PC connected via the USB port.

2. CI-V Address is the Icom address for the radio. The default is 94h. it should not be changed unless that is the only way that you can connect to a software package. Some earlier Icom radios used 88h and most radios don't use an address at all. The IC-7610 uses an address of 98h. You can change the address in the radio using the CI-V settings, but this would be a last resort because it would probably cause communication with other software that is expecting to use 94h to fail.

3. CI-V Transceive sets whether CI-V reports any change in status to the controlling PC program, or not. In the default ON mode, changing the VFO and other status settings is reported to the PC program. In the OFF position, the status data is only sent if the controlling PC program asks for it. Most PC software can only read data that has been specifically requested, so either setting should work fine.

4. CI-V USB→REMOTE is the address that is used for the RS-BA1 remote control software. Wayne Phillips has published a useful video at https://www.youtube.com/watch?v=pV4_xDtsMoY.

5. CI-V Output (for ANT) is used to select whether antenna control and frequency information are sent over the rear panel 'REMOTE' port. It is not relevant if you are using CI-V control over the USB cable.

6. CI-V USB Port. 'Link to [REMOTE]' causes CI-V commands to be sent to the REMOTE jack as well as the USB cable. 'Unlink from [REMOTE]' separates the Remote jack from the USB one. (See point 7).

7. The CI-V USB Baud Rate sets the data rate to the PC over the USB cable. But only if the CI-V USB Port is set to the default setting of 'Unlink from [REMOTE].'

8. CI-V USB Echo Back: If set to ON, the radio sends all received CI-V commands back to the software as a confirmation that they have been received. This is almost never required and if the PC software is also set to Echo ON, it can lead to problems. However, some versions of WSJT-X require Echo to be ON.

➢ **Audio Levels (USB cable)**

	Setting	IC-7300 default	ZL3DW setting	My Setting
	Windows 10 mixer input level to radio	N/A	28	
	Windows 10 mixer output level from radio	N/A	50	
1	AF Output Level	50%	50%	
2	USB MOD Level	50%	34%	
3	DATA OFF MOD	MIC, ACC	MIC, ACC	
4	DATA MOD	USB	USB	

1. Select <MENU> <SET> <Connectors> <ACC/USB AF Output Level>. AF output level is the audio level being sent to the PC over the USB cable. Since you can change the PC soundcard settings and often the digital mode software receive level, I chose to leave the AF output level set at the default setting of 50%.

2. Select <MENU> <SET> <Connectors> <USB MOD Level>. USB MOD level sets the level that will modulate the transmitter. Use the USB MOD control to set the transmit power while sending a digital mode from an external program. The ACC MOD Level control sets the transmit audio level from the ACC 1 jack.

3. Select <MENU> <SET> <Connectors> <DATA OFF MOD>. DATA OFF MOD sets the input source for the SSB, AM, and FM voice modes when DATA is turned off. It should be set to MIC or MIC & ACC.

4. Select <MENU> <SET> <Connectors> <DATA MOD>. DATA MOD sets the input source for the SSB-D, AM-D, and FM-D data modes when DATA is turned on. It should be set to USB, or ACC if you are using the ACC jack for audio, or MIC, USB if you are using the microphone connector for audio connections.

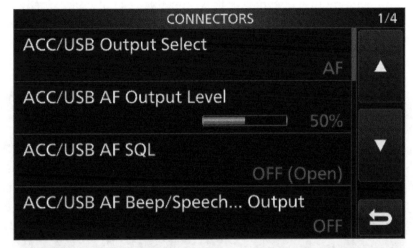

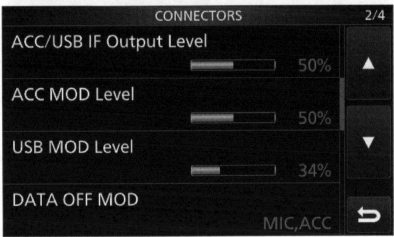

Audio settings

Function

The function button is the second button below the touch screen. It displays ten Soft Keys for enabling and disabling the; preamplifier, attenuator, AGC setting, manual and auto-notch, noise blanker, noise reduction, IP+, VOX, Compressor, ¼ tuning rate and transmit monitor. Several of these functions can also be turned on or off using front panel buttons or touch screen controls. A blue line around the icon indicates that the function is turned on or enabled. An amber line around the P.AMP/ ATT Soft Key indicates that the attenuator has been selected.

> **P.AMP/ATT Soft Key**

The P.AMP/ATT Soft Key controls the status of the internal preamplifier and front-end attenuator. The Soft Key has the same function as the front panel P.AMP/ATT button.

Touching the icon cycles through; Preamp Off, Preamp 1, and Preamp 2. A Blue outline indicates that a preamplifier is engaged.

Preamp 1 (≈7 dB) is recommended for the low HF bands. Preamp 2 (≈11-12 dB) is a higher gain amplifier recommended for the high HF bands. If signals are overloading the receiver, turn the preamplifier off.

Touch and hold turns enables a 20 dB attenuator. An amber outline indicates that the attenuator is in the circuit.

> **AGC Soft Key**

The AGC Soft Key cycles through the AGC fast, mid, and slow settings. Fast AGC is always selected for the FM mode.

Touching and holding the AGC Soft Key opens the AGC setting screen. To change an AGC time constant, select the operating mode SSB, AM, FM, RTTY, or a DATA mode using the normal mode selection icon. You can change radio modes without leaving the setup submenu. Then select FAST, MEDIUM, or SLOW. Use the Main VFO knob to adjust the value, not the Multi knob. Setting the time constant below 0.1 seconds turns the AGC off. Touch and hold the DEF icon to return the settings for the current radio mode to the Icom default. Press the EXIT button to close the window.

I have not changed any of these settings. I am sure that the default settings are appropriate. Unless AGC is off, a blue indicator is always displayed around the AGC Soft Key.

> **NOTCH function**

Touching the Notch icon cycles the notch filter, through auto, manual, and off.

The automatic notch filter (AN) and manual notch filter (MN) eliminate the effect of long-term interference signals such as carrier signals that are close to the wanted receiving frequency. Touch and holding the NOTCH Soft Key opens the MULTI menu where you can adjust the manual notch frequency and width. Start with a narrow notch and if it does not completely remove the interference signal, increase the width. The automatic notch filter will find an eliminate fixed carrier signals, "birdies."

> **NB (Noise Blanker)**

Touching the NB Soft Key turns the noise blanker on. The Soft Key has the same function as the NB button on the front panel.

Touch and hold the NB Soft Key to show the MULTI sub-menu where you can adjust the Level, Depth, and Width. The noise blanker is designed to reduce or eliminate regular pulse-type noise such as car ignition noise. You may need to experiment with the controls when tackling a particular noise problem.

The LEVEL control (default 50%) sets the audio level that the filter uses as a threshold. Most DSP noise blankers work by eliminating or modifying noise peaks that are above the average received signal level. They usually have no effect on noise pulses that are below the average speech level. Setting the NB level to an aggressive level may affect audio quality.

The DEPTH control (default 8) sets how much the noise pulse will be attenuated. Too high a setting could cause the speech to be attenuated when a noise spike is attacked by the blanker. This could cause a choppy sound to the audio.

The WIDTH control (default 50%) sets how long after the start of the pulse the output signal will remain attenuated. Set it to the minimum setting that adequately removes the interference. Very sharp short duration spikes will need less time than longer noise spikes such as lightning crashes.

The noise blanker is disabled when the radio is in FM mode.

Noise blanking occurs very early in the receiver DSP process. It is performed on the wideband spectrum before any demodulation or other filtering takes place. Noise reduction is performed on the filtered signal i.e. within the receiver passband.

➢ NR (Noise Reduction)

The noise reduction system in the IC-7300 is very effective. Touching the NR Soft Key has the same effect as pressing the NR button. Touch and hold the NR Soft Key to bring up a sub-menu where you can adjust the noise reduction level (default 5). Adjust the level to a point where the noise reduction is effective without affecting the wanted signal quality.

Noise reduction filters are aimed at wideband noise especially on the low bands rather than impulse noise which is managed by the noise blanker. The NR filter works best when the received signals have a good signal to noise ratio. By introducing a very small delay, the DSP noise reduction filter is able to look ahead and modify the digital data streams to remove noise and interference before you hear it. This is not easily achievable with analogue circuitry. It would require analogue delay lines.

➢ IP+ Soft Key

The IP+ Soft Key turns on the IP+ function. There is no touch and hold function. Turning IP+ on optimises the sampling system for best receive IMD (intermodulation distortion) performance. It can improve receiver performance in the presence of very large interfering signals. But there is a small loss of sensitivity. Turning IP+ off optimises the receiver's sensitivity.

The IP+ function turns on ADC randomisation and dither. While this may help during two-tone testing, there is normally enough band activity to make the use of ADC randomisation and dither unnecessary. So, in most cases, the IP+ function will have little or no noticeable effect and it can be left turned off. Adam Farson's test report shows a reduction of 3 dB in the MDS sensitivity with IP+ turned on.

Selecting IP+ makes no difference to the 'overflow' clipping level. Turning IP+ on created at least 8 dB of improvement in the DR_3 (3rd order IMD dynamic range) measurement. When IP+ mode is selected, a blue indicator is displayed around the IP+ Soft Key.

➢ **VOX Soft Key**

VOX stands for voice operated switch. When VOX is on, the radio will transmit when you talk into the microphone without you having to press the PTT button.

Using VOX is popular if you are using a headset or a desk microphone. Touching the VOX Soft Key turns the VOX on or off.

Touch and hold the VOX Soft Key to open the VOX setup MULTI submenu. Touch an icon to select the item to be adjusted. Turn the MULTI knob to change the setting. You should take the time to set the VOX up carefully as some settings tend to counteract other settings. In summary,

- VOX GAIN sets the sensitivity of the VOX, i.e. how loud you have to talk to operate it and put the radio into transmit mode. (Default 50%).

- ANTI-VOX stops the VOX triggering on miscellaneous noise like audio from the speaker or background noise. Higher values make the VOX less likely to trigger. (Default 50%).

- DELAY sets the pause time before the radio reverts to receive mode. It needs to be set so that the radio keeps transmitting while you are talking normally but returns to receive in a reasonable time when you have finished talking. (Default 0.2 seconds).

- VOICE DELAY sets the delay after you start talking before the transmitter starts. Generally, you want the radio to transmit immediately to avoid the first syllable or word being missed from the transmission. However, if you are prone to making noises perhaps you should set it longer. If you start each transmission with "Ah" you could set it quite long. (Default OFF).

➢ **COMP Soft Key**

The Compressor Soft Key is visible when the radio is set to one of the three 'voice' modes; AM, FM, or SSB. Touching the Soft Key turns the compressor on or off. A blue indicator around the Soft Key indicates that the compressor is on.

Touch and hold COMP to set the compressor level using the MULTI submenu. Touch the COMP icon to select and turn the MULTI knob to change the setting. (Default level = 5). See setting up the radio for SSB transmission, on page 9.

➤ **TBW (Transmit Bandwidth)**

Touch the TBW Soft Key to cycle through the three transmit bandwidth settings, WIDE, MID, or NAR. There is no touch and hold function.

WIDE is suitable for Rag Chewing on the local Net or chatting to locals on the 80m or 10m band. MID is better suited to working DX and the NAR (narrow) mode is suited to contest operation.

If you want to change the actual TBW bandwidth for each of the three options, see 'Setting the transmit bandwidth (TBW),' on page 11.

➤ **TONE Soft Key**

The TONE Soft Key is only visible when the transceiver is in FM mode. TONE is used for repeater operation primarily on the 6m band. The Soft Key has three modes; OFF, TONE, and TSQL. A blue indicator indicates that tone is in use.

- The TONE mode sends a tone with your FM transmission that opens the repeater squelch.

- The TSQL mode squelches your receiver until the correct tone is received from the repeater. It also sends the tone on your transmission to open the repeater squelch.

- The OFF mode turns off both receiver tone squelch and the transmitted tone.

Touch and hold the TONE key to open the setup screen.

- REPEATER TONE is the tone that is sent to the repeater in the TONE mode.

- T-SQL is the tone that must be received from the repeater to open the receiver squelch in the TSQL mode. The same tone is sent with your transmit signal.

The full set of CTCSS tones is available. Most repeaters use the default tone of 88.5 Hz or 67 Hz.

There is a useful icon called T-SCAN on the TONE FREQUENCY sub-menu.

- To find the tone that you should be transmitting for the TONE mode. Tune your receiver to the repeater input frequency then touch the REPEATER TONE icon. When another station is using the repeater, touch T-SCAN. The radio will scan through all the possible CTCSS codes until it finds the one that matches the tone on the other station's repeater input transmission. This method will probably work if the receiver is tuned to the repeater output frequency, as the same tone is usually transmitted by the repeater.

- To find the tone that you need for tone squelch in the TSQL mode. Tune your receiver to the repeater output frequency then touch the T-SQL - TONE icon. When another station is using the repeater, touch T-SCAN. The radio will scan through all the CTCSS codes until it finds the one that matches the tone on the repeater's output transmission.

Touch and hold DEF to reset the tone back to the default of 88.5 Hz.

> **MONI (Transmit Monitor)**

The MONI Soft Key turns on the audio monitor so that you can hear the signal that you are transmitting. Touch and hold the Soft Key to show the MULTI submenu. Turning the Multi knob adjusts the level of the monitor signal. A blue indicator around the Soft Key indicates that the transmit monitor is turned on.

'Monitor' is not available in the CW mode because sidetone is always on.

Unless you specifically want to listen to your transmit signal it is less distracting to leave the Monitor turned off. The SSB voice keyer will be heard if the Auto Monitor setting is 'on,' irrespective of the Monitor setting.

M.SCOPE

M.SCOPE toggles between a display with big VFO numbers and a small panadapter display. The small panadapter display has no Soft Keys so you can't change any of the panadapter settings unless you hold M.SCOPE to change to the big panadapter.

Depending on whether the 'EXPD/SET' button has been selected, holding M.SCOPE down for one second will change the display to either big VFO numbers with a small panadapter, or small VFO numbers with a bigger panadapter display.

M.SCOPE toggles between these two displays

Quick

The QUICK button is a bit of an oddity. It brings up a menu containing two items that could have been on the MULTI or FUNCTION menu buttons but are not. In fact, both of the options on the QUICK menu can also be accessed from other menu selections, so it seems that Icom just wanted to use the button for something.

The first item on the QUICK menu is METER TYPE. Selecting it opens a two-page sub-menu which lets you select the information that the meter displays when the transceiver is transmitting. This function is exactly the same as touching the meter scale on the display except that you go directly to the choice you want rather than having to cycle through the options.

The second item is <<REC Start>>. This starts recording audio to a file on the SD card. While a recording is in progress pressing QUICK again shows that the <<REC Start>> option has become <<REC Stop>>. Touching that stops the recording.

There are menu settings controlling what is recorded and for how long. See MENU RECORD on page 113. You can elect to record both your transmissions and the signal received off the air, or only the received signals. You can elect to record every transmission.

Exit

The EXIT button can be pressed to exit from any of the sub-menu displays including the MULTI sub-menus.

Rear panel connectors

REAR PANEL CONNECTORS

1.	DC 13.8 Volt power supply
2.	Ground Lug
3.	Antenna SO-239 jack (use a 50 Ω PL-259 plug)
4.	Cooling fan
5.	Icom antenna tuner - 4 pin Molex
6.	ALC (from linear amp) RCA white
7.	SEND (linear amp PTT) RCA red
8.	KEY (straight key or paddle) ¼" stereo phono
9.	ACC jack – 13 pin DIN
10.	USB port – USB 2.0 Type B
11.	Remote CI-V – mini phono (mono)
12.	External speaker B (Sub receiver) mini phono

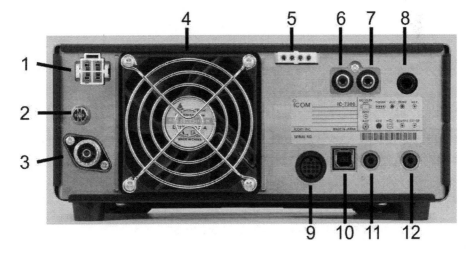

DC 13.8V

Connect the radio to a 13.8 Volt (± 15%) regulated power supply capable of supplying at least 21 Amps (23A recommended). Use the supplied power lead. **Never use a power lead without inline fuses.**

BE VERY CAREFUL TO SUPPLY THE CORRECT DC POLARITY. Red is positive and black is negative.

The connector has a locking tab. Squeeze the tab on the top of the connector down before attempting to unplug the DC cable.

GND

The radio ground connection should be connected to your shack ground in a 'star' rather than daisy chain format. The shack ground should be connected to an earth stake or earth mat outside. NOT to the mains earth. Earthing the radio can protect the radio from lightning static discharge (not lightning strikes). It can also improve noise performance.

ANT

The ANT connector is the primary antenna jack used for receiving and transmitting. The connector is a 50 Ω SO-239 UHF jack. It takes a PL259 plug.

COOLING FAN

Make sure that the fan is unobstructed.

TUNER

This 4-pin MOLEX jack is for connecting an Icom AH-4 or AH-470 antenna tuner.

ALC

The ALC and send jacks are used to control a non-Icom linear amplifier. Use a stereo RCA audio cable to connect the PTT and ALC lines to your linear amp. The IC-7300 ALC line accepts a voltage between 0 and -4 Volts with an input impedance of >3.3 kΩ. Check that the amplifier is compatible.

ALC should be configured so that it is not operating unless the transceiver is accidentally left at full power when driving the amplifier. Don't use it as a method of controlling the amplifier power. It should only be used as a failsafe in the event of a power setting mistake. ALC also protects the linear amplifier from transmitting overshoot. Some transceivers emit a full power RF spike when you key the transmitter even when the RF power is set to a low level.

SEND

The SEND jack is used as the PTT (transmit) signal for a non-Icom linear amplifier. You can set the delay between the time that the Send line goes low and RF is output from the antenna connector. Unless you are using a sequencer, or your linear amp includes one, it is strongly suggested that you set a delay of at least 10 ms to protect the amplifier relays. TX Delay: <MENU> <SET> <Function> <TX Delay>.

You can't use the SEND connector to key the transmitter. It is an 'output' only. The SEND pin on the ACC jack and the PTT pin on the Mic connector can be used to key the transmitter.

KEY

The rear panel KEY jack is for the connection of a CW bug, paddle, or straight key. It takes a standard ¼" (6.35 mm) stereo phono plug. The wiring is standard. Dit on the phono plug tip, Dah on the ring, and common on the sleeve. For a standard key use the tip and sleeve connections only.

You have to turn off the internal keyer if you want to use a straight key, bug, or external keyer. Select CW mode. There are eight menu settings under <MENU> <KEYER> <EDIT/SET> <CW-KEY SET>.

One setting that is not included in the KEYER EDIT/SET menu structure is the choice of having the CW signal on the transmitter's upper sideband or lower sideband. That is in <MENU> <SET> <Function> <CW Normal Side>. The default is LSB.

ACC

ACC (accessory jack) is a 13 pin DIN connector. It is intended for audio and control from a PC via an interface box. Something like an MFJ or RIGblaster interface. Other Icom radios have separate ACC1 (8-pin) and ACC2 (7-pin) accessory connectors. Icom dealers sell the OPC-599 conversion cable if you need to interface with equipment that requires the two-connector format.

You are more likely to use the USB connector which can do these functions with one cable. You might perhaps use an external interface box if you are operating in a multi-radio setup. The interface box should have audio transformers to isolate the receive and send audio signals and optocouplers or transistors to drive the RS232 control lines. See the Icom Basic Manual page 12-1 for more details.

Pin	Name	Function
1	8 V	Regulated 8 Volts ±0.3 V at a maximum of 10 mA. Used as a voltage reference for the band voltage output.
2	GND	Ground
3	SEND	Send / PTT input and output (goes low on transmit). If you are switching a relay, always put a reverse biased protection diode across the relay coil to protect the radio from back EMF. Input 'high' 2-20V. Input 'low' ground i.e. -0.5 to +0.8V. Output 'low' ground i.e. less than 0.1V max current 200mA.
4	-	Not used

5	BAND	Outputs a voltage between 0 and 8 Volts to indicate the band that the radio is transmitting on.
6	ALC	ALC input 0 to -4 Volts. Input impedance >3.3kΩ
7	-	Not used
8	13.8V	13.8 Volts at 1 Amp maximum
9	-	Not used
10	FSKK	FSK Keying for RTTY (FSK mode only) 'high' > 2.4 V, 'low' ground i.e. < 0.6 V, output current < 2 mA
11	MOD	Audio input to the radio, impedance 10 kΩ, level 100 mV
12	AF/IF	Audio, or 12 kHz IF output, from the radio, (menu choice). Level 100-300 mV at 4.7 kΩ
13	SQL S	Squelch output (goes low when the squelch opens). 'high' squelch closed >6.0 V 100 uA 'low' squelch open Ground i.e. < 0.3 V max 5 mA

USB PORT

The USB port is a USB 2.0 Type B port. You will need to purchase a 'Type A to Type B USB 2.0' cable to connect the radio to a PC.

USB is used for CI-V CAT control of the transceiver and for transferring audio to and from a PC for external digital mode software. It is also used to send decoded RTTY data to the PC and for remote control via the Icom RS-BA1 software.

To use the USB port, you need to download the USB driver software for your PC. See the 'Driver software' section on page 16. You will also have to set up the CI-V controls and the RTS and DTR keying lines, (both on page 118).

REMOTE

The REMOTE mono mini phone jack is used to connect the radio to a PC via an Icom CI-V to RS232 or USB adapter. It is there for compatibility with older radios and peripherals. If possible, use a USB cable connected to the USB port instead.

SPEAKER JACK

The speaker jack on the rear of the radio is for connection to an external speaker. It is a 1/8" (3.5mm) stereo jack (with a mono output).

Useful Tips

This chapter includes some techniques which may be useful, or at least interesting. I hate the default setting for the spectrum scope, so the first item in the chapter is a method of setting up the scope colours in a way that makes the display look like the panadapter display of most SDR receivers. The second item is a small paragraph about the screen-saver. I included it because I didn't even know there was a screen saver. The third item is a discussion on the mode switching. I can see why Icom configured it the way that they have, but I find it confusing. Considering that it is just a window on a touch screen and a bunch of Soft Keys I don't know why they didn't just add individual Soft Keys buttons for the SSB-D, AM-D, and FM-D data modes. The last item is about turning the AGC off. I don't recommend ever doing that even for digital modes, but I was intrigued to find that it is possible.

SETTING THE SPECTRUM VIEW TO A LINE VIEW

Unlike the audio spectrum display, the main spectrum scope display can't be set to a line rather than a filled in trace. I prefer to see a line, similar to the display on a spectrum analyser or almost every other SDR panadapter. Luckily careful adjustment of the screen colours allows you to fake it so that the filled in trace looks like a line display.

The spectrum can be set to show a filled spectrum or a filled spectrum with a line of a different colour on the top. There is also a max hold display and you can set the colour of that as well. Personally, I hate the max hold display and always turn it off.

To display a spectrum line instead of a filled spectrum, try the following settings.

1. These settings set the peak of the spectrum waveform to a white line and the filled in part of the display to a blue that is slightly darker than the waterfall giving the appearance of an unfilled spectrum trace.

2. Hold the M.SCOPE button for one second, until the large or medium panadapter is displayed.

3. Touch and hold the EXPD/SET Soft Key down for one second to display the settings.

4. Scroll to display 'Waveform Type' and set it to 'Fill+Line' then touch the Return icon ↺.

5. Touch 'Waveform Color (Current).' Write down the current colour settings in case you want to return to them later. Set the slider controls for 0 red, 0 green, and 57 blue. Then touch the Return icon ↺. You can set the colours back to default settings by touch and holding 'Waveform Color (Current).'

6. Touch 'Waveform Color (Line)' and set the controls for white. I have mine set a little less bright at, 198 red, 198 green, and 198 blue. Then touch the Return icon ↻. You can set the colours back to default settings by touch and holding 'Waveform Color (Line).'

7. I don't use the Max Hold indication, but if you do, you can probably leave the 'Waveform Color (Max Hold)' colours at the default, 58 red, 110 green, and 147 blue.

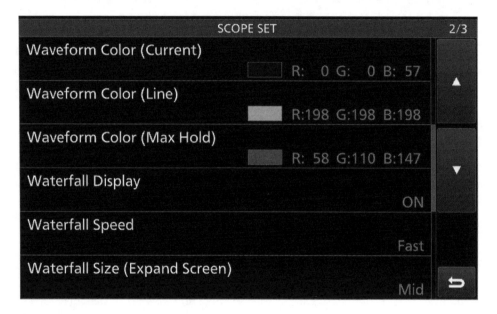

SCREEN SAVER MODE

The radio has a screen saver function to preserve your display just like your PC does. Touch the screen or press any button to restore normal operation. The green LED on the Power button flashes to indicate that the radio has gone into screen saver mode.

To restore the screen just press any button or touch the display screen.

You can turn the screen saver function off (not recommended) or set the time before the display goes to sleep using <MENU> <SET> <Display> <Screen Saver>. Note that received audio and the output to digital mode software via the USB cable will continue even though the screen is black.

RETURN TO SSB/FM/AM FROM A DATA MODE.

You would think that changing back to a voice mode from a data mode would be straightforward like selecting AM, FM, or SSB. But it does not work that way. The Data modes are considered to be sub-modes of the three voice modes. USB-D is a sub-mode of USB, FM-D is a sub-mode of FM, and AM-D is a sub-mode of AM.

- To return to USB from USB-D, don't select SSB mode. Touch DATA again.

- To return to LSB from LSB-D, don't select SSB mode. Touch DATA again.

- To return to FM from FM-D, don't select FM mode. Touch DATA again.

- To return to AM from AM-D, don't select AM mode. Touch DATA again.

ADJUSTING THE AGC TIME CONSTANT (OR TURN AGC OFF)

The AGC Soft Key on the FUNCTION menu cycles through Fast, Mid, and Slow.

Touch and holding the AGC Soft Key brings up a screen that allows you to change the AGC settings. The only way to exit this screen is to press the EXIT button.

- To adjust the time constant. Select the mode you want to change, Fast, Med, or Slow.

- Turn the main VFO knob to change the time constant. You can even set the AGC to Off. But other than for lab testing I don't know why you would want to do that. *SDR receivers have a much better dynamic range than Superheterodyne receivers, so weak signals should be unaffected by stronger signals nearby. AGC 'pumping' should not occur on the IC-7300.*

- You can change the operating mode without closing the AGC adjustment screen. Change the receiver mode and adjust the AGC settings for each mode in turn.

- The only way to exit the AGC adjustment screen is to press the EXIT button.

Troubleshooting

The items covered in the troubleshooting chapter are not faults. However, they are conditions that might worry you if you encounter them while operating the radio. This chapter may help you if something unexpected happens.

BLACK SCREEN AND A GREEN LED FLASHING ON THE POWER BUTTON

Don't panic! The radio has gone into screen saver mode. The radio has a screen saver function to preserve your display just like your PC does. Touch the screen or press any button to restore normal operation.

The green LED on the POWER button flashes to indicate that the radio has gone into screen saver mode.

You can turn the screensaver off, but I don't recommend doing so. You can change the time before the screen saver turns the screen off using <MENU> <SET> <Display> <Screen Saver>. [OFF, 15 min, 30 min, or 60 min].

WATERFALL TOO DARK, ONLY LARGE SPECTRUM PEAKS ARE SHOWING ON THE SPECTRUM DISPLAY.

Sadly, the panadapter REF level is the same for all bands, so if you change bands you are very likely to have to adjust the reference level.

If there are no Soft Keys at the bottom of the display press and hold the M.SCOPE button until they appear. Touch EXPD/SET to get the big panadapter. If there is no REF Soft Key at the bottom of the display, touch <1> to change the menu.

Touch the REF Soft Key and turn the main VFO knob clockwise to adjust the level until the spectrum noise floor is just visible at the above the waterfall. The waterfall brightness should now be correct. The reference level is indicated in a popup window. Touch REF again to exit the setup screen.

WATERFALL TOO LIGHT.

Sadly, the panadapter REF level is the same for all bands, so if you change bands you are very likely to have to adjust the reference level.

If there are no Soft Keys at the bottom of the display press and hold the M.SCOPE button until they appear. Touch EXPD/SET to get the big panadapter. If there is no REF Soft Key at the bottom of the display, touch <1> to change the menu.

Touch the REF Soft Key and turn the main VFO knob anti-clockwise to adjust the level until the spectrum noise floor is just visible at the above the waterfall. The waterfall brightness should now be correct. The reference level is indicated in a popup window. Touch REF again to exit the setup screen.

SMALL PANADAPTER HAS NO SOFT KEY CONTROLS.

The smallest panadapter has no Soft Key functions. Press and hold the M.SCOPE button until the medium size panadapter appears. Unfortunately, doing this will close the Keyer, Voice, or Decode window if it is open.

DATA MODE SELECTION CONFUSION

Data mode selection is initially a bit confusing. With the other modes, you just touch the mode you want. For example, to change from SSB to RTTY you touch RTTY. To change back to SSB you touch SSB. But this does not work if you touch DATA.

The Data modes are considered sub-modes of SSB, AM, or FM and the switching arrangement is different.

1. For a start, you can only select the DATA mode if the current mode is SSB, FM, or AM. If the current mode is PSK, RTTY, or CW, the DATA mode is not displayed. To get to the SSB-Data mode select USB or LSB first. To get to the FM-Data mode select FM first. To get to the AM-Data mode select AM first. Then touch DATA.

2. To get out of a DATA mode touch DATA, **not** SSB, AM, or FM.

TUNING RATE SPEEDS UP

The tuning rate may speed up as you turn the main VFO knob. This is normal. It is linked to the speed that you turn the knob. The feature is designed to get you to the other end of the band quickly when required. If you don't like it, you can turn it off using <MENU> <SET> <Function> <MAIN DIAL Auto TS>.

The tuning speed also changes when you are in 'fast mode' which has a 1 kHz step size. This is indicated by a small white triangle above the kHz number of the frequency display. Touch the three-number 'kHz' group on the VFO display to toggle between fast and normal mode.

Touch and hold the Hz digits of the VFO display to change between 10 Hz tuning steps to 1 Hz tuning steps. In data modes, RTTY and CW you can set a ¼ rate to slow down the main tuning even more.

SPECTRUM ON FIX DISPLAY MODE BUT NO MARKER

This can happen if you change band EDGEs on the panadapter, or use a memory channel, or change bands. It can also happen if you just tune off the end of the panadapter. The radio is able to operate on frequencies that are not currently being displayed on the panadapter spectrum scope display.

Two small green arrows ◀◀ at the left side of the panadapter indicate that the VFO is tuned to a frequency below the currently displayed spectrum. Two small green arrows ▶▶at the right side of the panadapter spectrum display indicate that the VFO is tuned to a frequency above the currently displayed spectrum.

NO SPECTRUM OR ONLY PART OF THE SPECTRUM DISPLAYED

This effect can happen if you have the panadapter in the FIX mode and you change bands or tune the VFO frequency off the edge of the currently selected fixed panadapter range. If the VFO frequency is so far from the fixed panadapter range that no spectrum can be displayed at all, a "Scope out of Range" warning is displayed. The problem will not occur if the panadapter is set to the CENT (centre) display type because with that setting the VFO frequency is always in the centre of the panadapter display. It can also be avoided by using the two scrolling panadapter modes Scroll-F in the FIXED display mode and Scroll-C in the CENT display mode.

VFO frequency too far from the range of the panadapter

The panadapter display can display a 1 MHz wide slice of the radio spectrum. The SDR software creates a spectrum display that is slightly over 2 MHz wide so that you can tune from side to side across the fixed 1 MHz panadapter window. If you tune more than 1 MHz either side of the fixed spectrum window, the spectrum display is truncated.

In the picture above, the panadapter EDGE is set to display frequencies between 28 and 29 MHz. If you tune within that range of frequencies, you can see the full spectrum and waterfall display. As you start to tune above 29 MHz the display shrinks. When the VFO is tuned to 29.550 MHz as shown in the picture, only about half of the spectrum remains visible.

To fix this, either select the CENT display type or return the VFO to a frequency within the panadapter display range. Or you may be able to touch the EDGE Soft Key to select a different FIX panadapter range. There are three panadapter band edges for each of thirteen frequency ranges. If you want to set them, see page 65.

TRANSCEIVER STUCK IN TRANSMIT MODE

➢ **Starting PC Digital Mode software causes the transceiver to go into transmit mode.**

This is not a fault, but it can be rather disconcerting. In SSB mode it is unlikely to damage anything as there will be very little power transmitted. But in CW AM or FM mode, it will cause power to be transmitted.

The problem is due to the RTS / DTS settings on the COM port being used by the radio being inappropriately controlled by the PC software. Normally the transceiver is set so that the RTS signal is PTT (turns transmit on) and the DTS signal is used to send CW or RTTY data. If the digital mode software has these controls reversed or either RTS or DTR is held 'active' or 'always on' it can make the transceiver switch to transmit. If this happens, check the COM port settings in the digital mode software and make sure that the RTS and DTR lines are set the same as the transceiver, or to 'always off.' This condition can be proved out of the radio by removing the USB cable.

It is often possible to set the digital mode software to use a CAT command for PTT rather than using the RTS (or DTR) line. But I use the RTS and DTR method.

> ➤ **Other reasons the transmitter may be stuck on.**

If it is not the Com RTS / DTR settings, check that the front panel TRANSMIT button has not been pressed. It could also be held on transmit by the Morse key or the microphone PTT. Check that the Mic PTT button is not stuck on transmit and temporarily remove the cables from the ACC jack and the KEY jack. The PTT pin on the ACC jack also has the capability to turn on the transmitter. Basically, unplug all external cables except the antenna and the DC supply until the problem is resolved.

CW MODE SIDETONE BUT NO TRANSMIT POWER

This one has caught me out several times. The CW mode won't automatically transmit unless full or semi break-in has been selected. In CW mode press the VOX/BK-IN button and choose either BKIN (semi break-in) or F-BKIN (full break-in) mode and the radio will transmit. Alternatively, with BKIN set to OFF, you can key the PTT line by pressing TRANSMIT, pressing the PTT button on the microphone, via a CI-V command, or by switching the SEND line on the ACC jack.

The following settings affect the sending of keying macro messages as well.

- With BK-IN OFF you can practice CW by listening to the side-tone without transmitting.

- Full break-in mode F-BKIN will key the transmitter while the CW is being sent and will return to receive as soon as the key is released. This allows for reception of a signal between CW characters. Unfortunately, the IC-7300 uses a relay for PTT switching, so full break-in is quite noisy.

- Semi break-in mode BKIN will key the transmitter while the CW is being sent and will return to receive after a delay when the key is released. In CW mode, touch and hold the BKIN Soft Key to adjust the delay. Turn the Multi knob to change the setting. The default is a period of 7.5 dits at the selected KEY SPEED.

OVF ALERT

The OVF alert is displayed at the top of the main screen to the right of the FIL (filter) indicator. It indicates that the receiver ADC is being overloaded. Turn off the preamplifiers and if necessary, turn on the attenuator using the P.AMP ATT button. Or you can turn down the RF gain by turning the RF/SQL knob anti-clockwise.

Note that the signal that is causing the overload may not be the signal that you are listening to and may not necessarily be visible on the panadapter spectrum display. Adding attenuation using the ATT Soft Key should fix the problem.

When receiving medium wave AM broadcast stations below 1.7 MHz, interference from strong stations can be mitigated by turning the MF Band ATT on. Select <MENU> <SET> <Function> <MF Band ATT> <ON>. (Default ON).

The MF Band attenuator is a 16 dB attenuator which is inserted into the receive chain when the radio is tuned below 1.7 MHz. If you want to receive weak MF stations or frequencies below 1.7 MHz you can turn the attenuator off provided you reduce the RF gain to a point where the OVF indicator remains off.

LMT ALERT

The LMT icon is below the red TX it indicates that the transceiver has overheated while transmitting and the protection system has activated. The TX Icon will also turn to white lettering on a black background.

PROBLEMS TRANSMITTING DIGITAL MODES

The audio signal is sent over the USB cable to the PC in any mode. So, you can use your digital mode PC software to see and decode digital mode signals like RTTY, PSK, or FT8. But you can't transmit audio digital mode signals unless you are in a DATA mode, usually USB-D. Use FM-D for FM Packet radio on 6m.

CW is digitally keyed, so an external program can key CW when the transceiver is in CW mode. AFSK RTTY is transmitted using the SSB-D mode. FSK RTTY is transmitted with the radio in RTTY mode.

VOICE KEYER, CW, OR RTTY MESSAGE KEYER NOT WORKING

Check that the message keyer matches the mode. They look right but they won't work. Check the text above the eight message buttons. VOICE TX for voice modes, KEYER for CW, and RTTY DECODE for RTTY.

The Soft Keys are T1-T8 for Voice, M1-M8 for CW, and RT1-RT8 for RTTY.

HELP, I SET MY RADIO TO JAPANESE TEXT MODE

If you set the radio to display Japanese text and can't find your way back. It probably serves you right!

Select <MENU> <SET> Then touch the ディスプレイ設定 'Display Setting' icon, second to bottom on the left side. It is the one with a picture of a computer screen on it.

Then select 表示言語 'Display Language' It is the bottom menu item on page 4/4.

Finally, select 'English' 英語. It is the top menu item of the two options.

KEYBOARD AND COMPUTER MOUSE SUPPORT

Some forum users have been wondering how to connect an external mouse or keyboard to the IC-7300. The sad fact is that you can't. The radio only has one USB port and it is a type B port.

A USB Type B port is used when the device, in this case, the transceiver, is a peripheral connected to a remote host, the PC. It is used on items like scanners or printers. A USB Type A port is a host port used for connecting a USB peripheral such as a keyboard, a mouse or a USB memory stick.

Unlike its big brother the IC-7610, the IC-7300 does not have a USB type-A port so you cannot connect a mouse or a keyboard. This is a big problem if you want to use the built-in RTTY functions and to a lesser degree the CW message functions. The problem is that there is no way to input the callsign of the station you are working or giving a signal report to. You can't type a message either. It makes the built-in RTTY function 90% useless. I guess you could edit the RTTY message using the on-screen keyboard each time you want to send someone's callsign, but I don't think that is very practical during a QSO. Perhaps you could edit the message and then call the station. Unfortunately, there is no way to fix the problem because it is not a software issue. The radio simply does not have the USB type A hardware.

CLOCK LOSES TIME OR RESETS

The IC-7300 has a rechargeable 3 Volt battery to keep the clock working while the radio is disconnected from the 13.8 V DC supply. If you leave the radio disconnected for an extended period, the battery will discharge, and the clock will reset. If the voltage gets too low the battery will be damaged and require replacement which is a solder job as it is not in a holder. Note that if the radio is left connected to your DC supply the clock battery will charge even if the radio is turned off. The obvious solution is to leave your DC supply running and turn the rig off with the power button. Apparently, it can take up to two days to charge from a fully discharged state.

PANADAPTER GAIN

This really bugs me. No, you can't change the gain of the panadapter band scope display. You can change the reference level, which shifts the whole panadapter up or down the screen, but the gain is set to 10dB per division. This means that a lot of the time the panadapter will display tiny signal peaks just above the noise floor. Adding a control to adjust the panadapter gain or a Soft Key on the EXPD/SET to set a choice of 5 dB or 10 dB per division would be a big improvement.

Modifications

Please note that I am not endorsing any modification to the radio and that carrying out any modification will almost certainly void the Icom warranty.

"This warranty shall not apply: To an Icom product which has failed to function as required due to improper installation, misuse, accident, alteration or unauthorised repair or modification."
[RWB Communications website – NZ Icom agent].

LOW AVERAGE TRANSMITTER POWER

Some people have complained about the average transmitter output power being low. And some people say that the modulation power is deliberately reduced so that the transceiver achieves good transmitter linearity. I believe that this perceived low transmitter power problem is due to not setting the microphone and compression levels correctly. See Setting up the radio for SSB operation on page 9.

Also, the very fast power meter on the Icom display can give the illusion that the transmitter is peaking to full power but under-delivering on average power.

Personally, I believe that a clean signal is more important than a loud signal. Remember that a reduction of average power from 70 Watts to 35 Watts is only a 3 dB decrease in the signal received at the distant station. That is only half of one S point on the receiver's S meter. In other words, this is a "non-problem."

- Low average transmit power may be due to lack of bass in your modulating audio. Maybe the low cut in your transmit bandwidth (TBW) is set too high. For example, the low cut on the TBW narrow setting could be at 500Hz. Try setting TBW to MID or WIDE.

- You could also look at the audio scope while transmitting with TBW set to WIDE. Observe the frequency response of the transmission on the audio spectrum scope. It should be reasonably flat. You can tweak the TX Bass and TX Treble controls on the <MENU> <SET> <Tone Control/TBW> <TX> <SSB> menu tab.

- Forget about average power. If you make sure that ALC never goes above 50% and that COMP is averaging 10% to 20%, your transmission will sound good.

If you are still unhappy with your transmitter output power and you have set the audio levels as specified in 'Setting up the radio for SSB operation on page 9,' there is a hardware modification. I **really** do not think that it is necessary.
http://sp3rnz.blogspot.com/2017/01/icom-ic-7300-ssb-power-mod.html

ANTENNA TUNER MODIFICATION BY ALEX SP9SOY

http://radioaficion.com/cms/ic-7300-antenna-tuner-range-modification/

This modification extends the ability of the internal antenna tuner to match mismatched antenna loads. Please note that the reason that Icom has restricted the tuning range is to avoid the possibility of overheating or overvoltage damage to the tuner components. Therefore, although the modification may allow you to tune a mismatched antenna, you do run the risk of damaging the radio.

ICOM IC-7300 WIDE BAND MODIFICATION BY PA2DB

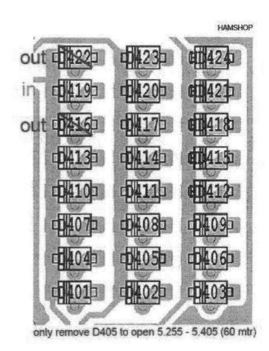

only remove D405 to open 5.255 - 5.405 (60 mtr)

Please note that this modification requires the removal of very small SMD diodes (surface mount devices). This is delicate work that should not be attempted by anyone that is inexperienced with SMD devices and delicate soldering.

If you are reinstalling diodes make sure they are the right way around.

I have no need to change the parameters of the radio, so I have not tried this mod.

1. To open the receiver from 0.030 MHz to 74.8 MHz remove diode D416. My receiver already covers that range, so I don't need to make this change.

2. To open the 60m band, 5.255 to 5.405 MHz remove diode D405. Again, my transceiver already covers that range and I can transmit on the 60m band, so I don't need to make this change.

3. To open the transmitter so that it can transmit over the entire 0.030 MHz to 74.8 MHz range, for MARS or Marine Band operation etc. Remove D422.

4. Do not remove D419.

The diode drawing is from http://radioaficion.com/cms/ic-7300-wide-band-modification/

The diode photo is by Jörg Saure from Google – the link to the original site is dead.

FAN MODIFICATION

The fan upgrade is the only modification for the IC-7300 that I believe is a really good idea. The stock fan on the IC-7300 is quite noisy and it starts at high speed every time you transmit, turning its speed down when it has decided that the radio isn't hot. I find it rather annoying when transmitting FT8 as it comes on with a rush for every 15-second transmission.

The modification is non-intrusive and could be easily reversed if you had to send the radio to an Icom service agent.

Bjorn Eklund, SM7IUN and Pete Robinson G4IZH recommend replacing the stock fan with a Noctua NF-A8 PWM fan. Make sure it is the 12 Volt model. I don't believe it has to be a PWM model as the two PWM (pulse width modulated) control wires are not used. These fans are specially designed to be exceptionally quiet. The replacement fan is essentially silent. A big improvement over the stock fan. http://www.g4izh.co.uk/noisy-fan-mod.html

TRANSVERTER INTERFACE KIT

Kuhne Electronic sells a transverter interface kit for the IC-7300 SDR Transceiver. The small PCB replaces the Molex connector that would be used to connect an Icom specific tuner. The Molex connector can be tucked away inside the radio, so that it can be reinstalled if you need to remove the Transverter board.

See https://shop.kuhne-electronic.com.

Kuhne transverter interface kit

Photo: http://rdp.cat/contingut/icom-ic-7300-mods/

INRAD RECEIVER ANTENNA PORT MODIFICATION

The INRAD model RX7300 allows you to add a receive-only antenna jack to the Icom IC-7300. It is an easy to install and easily removable plug-in modification that requires no soldering. Available from INRAD and a range of radio dealers.

http://www.inrad.net/product.php?productid=371

INRAD receive antenna mod.

Photo: http://www.inrad.net/product.php?productid=371

Glossary

Term	Description
6m, 10m 20m, 40m, 80m	50 MHz, 28 MHz, 14 MHz, 7 MHz, and 3.5 MHz amateur radio bands
59	Standard (default) signal report for amateur radio voice conversations. A report of '59' means perfect readability and strength.
599, 5NN	Standard (default) signal report for amateur radio CW conversations. A 599 report means perfect readability, strength, and tone. The 599 signal report is often used for digital modes as well. The 5NN version is faster to send using CW. It is often used as a signal report exchange when working contest stations.
73	Morse code abbreviation 'best wishes, see you later.' It is used when you have finished transmitting at the end of the conversation.
.dll	Dynamic link library. A reusable software block which can be called from other programs.
A/D	Analogue to digital
ACC	Accessory jack
ADC	Analog to digital converter or analog to digital conversion
AF	Audio frequency - nominally 20 to 20,000 Hz.
AFSK	Audio Frequency Shift Keying. RTTY mode that uses tones rather than a digital signal to drive the SSB transmitter.
AGC	Automatic Gain Control. In an SDR like the IC-7300, it limits the receiver audio output in the presence of large signals.
ALC	Automatic Level Control. There are two kinds used by the radio. One is the ALC used to ensure that the radio is not overmodulated. It is metered by the ALC meter. The other is the ALC control from a linear amplifier which ensures that it is never overdriven by the transceiver.
Algorithm	A process, or set of rules, to be followed in calculations or other problem-solving operations, especially by a computer. In DSP it is a mathematical formula, code block, or process that acts on the data signal stream to perform a particular function, for example, a noise filter.
AM	Amplitude modulation, (double sideband with carrier)

ANF	The Automatic Notch Filter eliminates the effect of long-term interference signals such as carrier signals that are close to the wanted receiving frequency. Not effective against impulse noise.
ANT	Abbreviation for antenna
APF	Audio Peak Filter. This is a very sharp filter (3 bandwidth settings) that operates on the DSP audio signal. It is only used in the CW mode.
ATT	Abbreviation for attenuator
Auto Tune	This is a button which pulls the radio VFO on to the frequency that matches the CW Pitch that you have set. It 'Nets' the CW receive frequency and transmit frequency so that you will transmit CW on the same frequency as you are receiving. It only works in the CW mode.
Band scope	A band scope is a spectrum display of the frequencies above and below the frequency that the radio is tuned to. The centre of the display is generally the frequency that you are listening to. This is different from a panadapter where you can listen to any frequency across the display.
Bit	Binary value 0 or 1.
BKIN or F-BKIN	CW 'Break-in' the practice of receiving Morse code between the Morse characters or words that you are sending. The radio will not automatically transmit in the CW mode unless BKIN is set to either semi break-in BKIN or full break-in F-BKIN.
BPSK	Binary phase shift keying. Digital transmission mode using a 180-degree phase change to indicate the transition from a binary one to a binary zero. The number is the baud rate. BPSK31 is slower but uses less bandwidth and is easier to decode than BPSK63.
Carrier	Usually refers to the transmission of an unmodulated RF signal. It is called a carrier because the modulation process modifies the un-modulated RF signal to carry the modulation information. A carrier signal can be amplitude, frequency, and/or phase modulated. Then it is referred to as a 'modulated carrier.' An oscillator signal is not a carrier unless it is transmitted.
CAT	Computer-aided transceiver. Text strings used to control a ham radio transceiver from a computer program. The Icom standard for CAT control is called CI-V
CENT	Centre mode. Sets the panadapter display so that the main VFO frequency is in the centre of the display with a SPAN of frequencies either side.

CI-V	Icom standard for CAT control. Text strings used to control a ham radio transceiver from a computer program
CODEC	Coder/decoder - a device or software used for encoding or decoding a digital data stream.
COM	Serial Communications port. In the IC-7300 the Com Port is a 'virtual' serial port carried over the USB cable. The dedicated CI-V 'REMOTE' connector is also a COM port but a level converter is required.
COMP	Compressor. Increases the average power of your transmission by decreasing the dynamic range of the audio signal. i.e. makes the quiet parts louder.
CPU	Central processing unit - usually a microprocessor. Can be implemented within an FPGA.
CQ	"Seek You" an abbreviation used by amateur radio operators when making a general call which anyone can answer.
CW	Continuous Wave. The mode used to send Morse Code.
D/A	Digital to analogue.
DAC	Digital to analog converter or digital to analog conversion
DATA	One of the data modes (AM-D, FM-D or SSB-D) used to interface the radio with a PC digital mode program. You must be in a DATA mode to transmit from a PC digital mode program.
dB, dBm, dBc, dBV	The Decibel is a way of representing numbers using a logarithmic scale. dB is used to describe a ratio, i.e. the difference between two levels or numbers. Decibels are often referenced to a fixed value such as a Volt (dBV), a milliwatt (dBm), or the carrier level (dBc). Decibels are also used to represent logarithmic units of gain or loss. An amplifier might have 3 dB gain. An attenuator might have 10 dB loss.
DC	Direct Current. You need a 23 Amp regulated 13.8V DC supply to power the radio.
Digital modes	Amateur radio transmission of digital information rather than voice. It can be text, or data such as video, still pictures, or computer files, (PSK, RTTY, FT8, Olivia, SSTV etc.)
Digital voice	Amateur radio mode where speech is coded into a digital format and sent as tone sequences or phase shift keying.
DR3	3rd order dynamic range receiver test

DSP	Digital signal processing – a dedicated integrated circuit chip usually running internal firmware code, or a software program running on a computer. DSP uses mathematical algorithms in computer software to manipulate digital signals in ways that are equivalent to functions performed on analogue signals by hardware mixers, oscillators, filters, amplifiers, attenuators, modulators, or demodulators.
DTR	'Device terminal Ready' a com port control line often used for sending CW or FSK RTTY data over the CI-V interface between the radio and a PC.
DX	Long distance, or rare, or wanted by you, amateur radio station
DXCC	The DX Century Club. An awards programme based around confirming contacts with 100 DXCC 'countries' or 'entities' on various modes and bands. The DXCC list of 304 currently acceptable DXCC entities is used as the worldwide standard for what is a separate country or recognisably separate island or geographic region.
DXpedition	A DXpedition is a single, or group of, amateur radio operators who travel to a rare or difficult to contact location, for the purpose of making contacts with as many amateur radio operators as possible worldwide. They often activate rare DXCC entities or islands and they may operate stations on a number of bands and modes simultaneously.
EXPD	Abbreviation for Expanded
EXT	Abbreviation for External
FFT	Fast Fourier Transformation – conversion of signals from the time domain to the frequency domain (and back using IFFT).
FIX	Fixed mode. Sets the panadapter display so that it displays the range frequencies between two pre-set frequencies determined by Band Edges. The VFO frequency is indicated by the green marker.
FM	Frequency Modulation. Used for repeaters on the 6m band.
FPGA	Field Programmable Gate Array – a chip that can be programmed to act like logic circuits, memory, or a CPU.
FSK	Frequency Shift Keying.
FSK RTTY	FSK RTTY is keyed using a digital signal to offset the transmit frequency rather than AFSK which generates audio tones at the Mark and Space offsets from the VFO frequency.
GND	Ground. The earthing terminal for the radio. This should be connected to a 'telecommunications' ground spike, not the mains earth.

GPS	Global Positioning System. A network of satellites used for navigation, location, and very accurate time signals.
GPSDO	GPS disciplined oscillator. An oscillator locked to time signals received from GPS satellites
Hex	Hexadecimal – a base 16 number system used as a convenient way to represent binary numbers. The default Icom address for the IC-7300 is 94h (148 decimal) (10010100 binary)
HF	High Frequency (3 MHz -30 MHz)
Hz	Hertz is a unit of frequency. 1 Hz = 1 cycle per second.
IF or I.F.	Intermediate frequency = the Signal – LO (or Signal + LO) output of a mixer
IMD	Intermodulation distortion. Interference or distortion caused by non-linear devices like mixers. There are IMD tests for receivers and transmitters. IMD performance of linear amplifiers can also be tested.
IP+	The IP+ control enables ADC randomisation and 'dither' in order to optimise the sampling system for best receive IMD (intermodulation distortion) performance. It can improve receiver performance in the presence of very large interfering signals. But there is a small loss of sensitivity. While it improves laboratory 'two tone IMD test' results, it is usually unnecessary in 'on air' situations.
IQ	Refers to the I and Q data streams treated as a pair of signals. For example, a digital signal carrying both the I (incident) and Q (quadrature) data.
Key	A straight key, paddle, or bug, used to send Morse Code
kHz	A kilohertz is a unit of frequency. 1 kHz = 1 thousand cycles per second.
LAN	Local Area Network. The Ethernet and WIFI connected devices connected to an ADSL or fibre router at your house is a LAN.
LED	Light Emitting Diode
LSB	Lower sideband SSB transmission
m, 2m, 6m	Meter (US) or Metre. Often used to denote an amateur radio or shortwave band; e.g. 2m, 6m, 30m, 10m, where it denotes the approximate free space wavelength of the radio frequency. Wavelength = 300 / frequency in MHz. A frequency range of 3 to 30 MHz has a corresponding wavelength of 100m to 10m.
Marker	The R and T markers indicate the frequency of the receiver (in FIX mode) and the transmitter frequency (press XFC).

MDS	Minimum discernible signal. A measurement of receiver sensitivity
Menu	The MENU button displays a wide range of settings that are not used enough to warrant a dedicated front panel button.
MHz	Megahertz – unit of frequency = 1 million cycles per second.
MIC	Microphone
MPAD	Memory Pad. A short-term memory function, storing either five or ten frequency and mode settings
MW	Memory Write Soft Key
NB	Noise Blanker. A filter used to eliminate impulse noise
Net	An on-air meeting of a group of amateur operators. Or to 'Net' the CW receive frequency and transmit frequency so that you will transmit CW on the same frequency as you are receiving
NR	Noise Reduction. A filter used to eliminate continuous background noise
Onboard	A feature performed within the radio. Especially one that usually requires external software. For example, the radio has an 'onboard' RTTY decoder.
OVF	Overflow. The overflow warning indicates that the receiver ADC is being overloaded with very big received signals. This will cause problems on the panadapter display and possibly create noise due to severe intermodulation distortion. Turn on the attenuation with the ATT button or reduce the RF Gain (RF/SQL control).
P.AMP	An abbreviation for 'preamplifier'
Panadapter	Panadapter is short for Panoramic Adapter. It allows us to see a panoramic display of the band. The IC-7300 panadapter can display a spectrum display and optionally a waterfall picture. You may be listening to one or more signals anywhere within the displayed spectrum of frequencies. This is different from a band scope.
PC	Personal Computer. For the examples throughout this book, it means a computer running Windows 10.
Pileup	A pileup is a situation when a large number of stations are trying to work a single station, for example, a DXpedition or a rare DXCC entity. Split operation is often employed to spread the pileup of calling stations over a range of frequencies.
Po	RF power output (meter)
PROG	PROG (Programmed) is a Soft Key used to scan between the pre-set P1 and P2 frequencies.

PSK	Phase shift keying. Digital transmission mode using phase change to indicate the transition from a binary one to a binary zero.
PTT	Press to talk - the transmit button on a microphone – The PTT signal sets the radio and software to transmit mode. Icom calls it 'SEND.'
QPSK	Quadrature phase shift keying. Digital transmission mode using 90-degree phase changes to indicate four two-bit binary states 00,01,10,11.
QRP	Q code - low power operation (usually less than 10 Watts).
QSO	Q code – an amateur radio conversation or "contact."
QSK	Q code – fast transmit to receive switching which allows Morse code to be received in the gaps between the CW characters that you are sending.
QSY	Q code – a request or decision to change to another frequency.
RBW	Resolution Bandwidth is the ability of the spectrum scope (panadapter) to distinguish between signals that are on frequencies that are very close together. A narrow RBW can display signals that are closer together but requires more processing power and computer processing speed.
REC	Record. A button used to record off-air received audio. Also, a Soft Key used to record microphone audio for the Voice Message keyer.
REF	Reference
RF	Radio Frequency
RIT	Receive Incremental Tuning. A way to fine tune the signal that you are receiving without changing the main VFO and hence your transmitted frequency. Used when the other station is a little off frequency.
RS232	A computer interface used for serial data communications.
RTTY	Radio Teletype. RTTY is a frequency shift, digital mode. Characters are sent using sequences of Mark and Space tones.
RTS	'Ready to Send' a com port control line often used for sending the PTT (SEND) command over the CI-V interface between the radio and a PC.
RX	Abbreviation for receiver
SDR	Software Defined Radio. Actually, the IC-7300 is more correctly a 'direct sampling' radio.
Sked	A pre-organised or scheduled appointment to communicate with another amateur radio operator

SNR	Signal-to-Noise Ratio in dB (decibels).
Soft Key	A button or selectable icon displayed on the touch screen
Split	The practice of transmitting on a different frequency to the one that you are receiving on. Split operation is commonly used by DXpeditions and anyone who generates a large pileup of callers. SSB split is commonly 5-10 kHz. CW split is commonly 1-2 kHz.
Squelch	Squelch mutes the audio to the speakers when you are not receiving a wanted signal. When the received signal level increases the squelch opens and you can hear the station. Squelch does not affect the audio output over the USB cable or the audio meter display, (menu setting).
SSB	Single sideband transmission mode.
SWR	Standing Wave Ratio. The RF power reflected back from a mismatched antenna or connection.
TX	Abbreviation for Transmit or Transmitter
UHF	Ultra High Frequency (300 MHz - 3000 MHz).
USB	Universal serial bus – serial data communications between a computer and other devices. USB 2.0 is fast, USB 3.0 is very fast.
USB	Upper sideband SSB transmission.
UTC	Universal coordinated time. UTC is the standard time used by amateur radio operators. Everyone logging contacts using the same UTC time rather than local time makes comparing logs and confirming contacts much easier.
VBW	Video Bandwidth is the ability of the spectrum scope (panadapter) to distinguish weak signals from noise. A narrow VBW can filter noise but requires more processing power.
VFO	Variable Frequency Oscillator. The IC-7300 has two VFOs called 'A' and 'B.' The main tuning knob controls the active VFO.
VHF	Very High Frequency (30 MHz -300 MHz)
VOX	Voice Operated Switch. Voice-activated receive to transmit switching
W	Watts – unit of power (electrical or RF).
XIT or ⊿TX	Transmit Incremental Tuning. A way to fine-tune your transmitted signal without changing the main VFO and hence your receiver frequency.

Index

References and Links

1. Icom IC-7300 Basic Manual (copyright Icom)

2. Icom IC-7300 Full Manual (copyright Icom)

3. There is an excellent video on setting up the Icom RS-BA1 remote control software at https://www.youtube.com/watch?v=vhhNo2AoO0Y

4. Icom website.
   http://www.icom.co.jp/world/support/download/firm/index.html.

5. Radio Society of Great Britain www.rsgb.org

6. Notes on FSK RTTY by K0PIR at http://www.k0pir.us/icom-7300-rtty-fsk-mmtty/ or a video on the topic at https://www.youtube.com/watch?v=NmNHVjjAdiY.

7. Eterlogic VSPE (Virtual Serial Port Emulator) software. See http://www.cedrickjohnson.com/2017/03/icom-ic-7300-usb-for-radio-control-fsk-keying/ for Cedrick Johnson's explanation.

8. Icom band output decoders
   http://kk1l.com/kk1l_2x6switch/KK1LIcomDecoderInstructions2d2.pdf.

   https://remoteqth.com/arduino-band-decoder.php.

   http://www.k6xx.com/radio/icbsciv.pdf

9. CI-V REMOTE jack https://www.youtube.com/watch?v=pV4_xDtsMoY.

10. Modifications

    http://www.inrad.net/product.php?productid=371

    http://radioaficion.com/cms/ic-7300-antenna-tuner-range-modification/

    http://radioaficion.com/cms/ic-7300-wide-band-modification/

    http://www.g4izh.co.uk/noisy-fan-mod.html

    https://shop.kuhne-electronic.com.

    http://www.inrad.net/product.php?productid=371

The Author

Well if you have managed to get this far you deserve a cup of tea and a chocolate biscuit. It is not easy digesting large chunks of technical information. It is probably better to use the book as a technical reference. Anyway, I hope you enjoyed the book and that it has made life with the IC-7300 a little easier.

I live in Christchurch, New Zealand. I am married to Carol who is very understanding and tolerant of my obsession with amateur radio. She describes my efforts as "Andrew playing around with radios." We have two children and a cat. One son has recently graduated from Canterbury University with a degree in Commerce and the other is studying Medicine at Otago University. The cat stays mostly at home.

I am a keen amateur radio operator who enjoys radio contesting, chasing DX, digital modes, and satellite operating. But I am rubbish at sending and receiving Morse code. I write extensively about many aspects of the amateur radio hobby, writing regular columns for Break-in and Radio User magazines. This is my fifth book.

Thanks for reading my book!

73 de Andrew ZL3DW.

THE END
73 and GD DX

Quick Reference Guide

Antenna tuner	\<MENU> \<SET> \<Function> \<TUNER>
Audio input level from PC (USB cable)	\<MENU> \<SET> \<Connectors> \<USB MOD Level>
Audio input level from PC (ACC jack)	\<MENU> \<SET> \<Connectors> \<ACC MOD Level>
Audio output level to PC (USB cable or ACC jack)	\<MENU> \<SET> \<Connectors> \<ACC/USB AF Output Level>
Audio or 12 kHz IF output to USB cable or ACC jack	\<MENU> \<SET> \<Connectors> \<Output Select>
Audio scope settings	\<MENU> \<AUDIO> \<hold EXPD/SET>
Band Edge (FIX panadapter)	Hold \<M.SCOPE> Hold \<EXPD/SET> \<Fixed Edges>
Band Edge (radio) setting	\<MENU> \<SET> \<Function> \<User Band Edge> (change beep setting to enable)
Beep settings	\<MENU> \<SET> \<Function> \<Band Edge Beep>
Callsign on power on splash screen	\<MENU> \<SET> \<Display> \<My Call>
CI-V (CAT) settings	\<MENU> \<SET> \<Connectors> \<CI-V>
Clock – set date and time	\<MENU> \<SET> \<Time Set>
Compressor	\<FUNCTION> \<COMP>
CW Key / paddle	\<MENU> \<KEYER> \<EDIT/SET> \<CW-KEY SET> \<Key Type>
CW messages (CW mode)	\<MENU> \<KEYER> \<EDIT/SET> \<EDIT>
CW messages via keypad F1 - F8	\<MENU> \<SET> \<Connectors> \<External Keypad> \<KEYER>
CW settings	\<MENU> \<KEYER> \<EDIT/SET>
CW sideband LSB (default)	\<MENU> \<SET> \<Function> \<CW Normal Side>
CW sidetone (CW mode)	\<MENU> \<KEYER> \<EDIT/SET> \<Sidetone Level>
Echo on (for WSJT-X FT8)	\<MENU> \<SET> \<Connectors> \<CI-V> \<CI-V USB Echo Back> \<ON>
Firmware information	\<MENU> \<SET> \<Others> \<Information>
Firmware update (SD card)	\<MENU> \<SET> \<SD Card> \<Firmware Update>
FM Split offset (6m Band)	\<MENU> \<SET> \<Function> \<SPLIT> \<FM SPLIT Offset (50M)>

FM Split offset (HF Bands)	\<MENU\> \<SET\> \<Function\> \<SPLIT\> \<FM SPLIT Offset (HF)\>
High and Low pass filters (RX)	\<MENU\> \<SET\> \<Tone Control/TBW\> \<RX\> *\<mode\>* \<RX HPF/LPF\>
Lock (VFO or panel dial lock)	\<MENU\> \<SET\> \<Function\> \<Lock Function\>
Memory	\<MENU\> \<MEMORY\>
Memory Pad	\<MENU\> \<MPAD\>
Memory Pad size (5 or 10)	\<MENU\> \<SET\> \<Function\> \<Memo Pad Quantity\>
Mic Gain (SSB, AM, FM)	\<MULTI\> \<Mic Gain\>
MF Band Attenuator	\<MENU\> \<SET\> \<Function\> \<MF Band ATT\>
Monitor (transmit)	\<MULTI\> \<MONITOR\>
Multi-function meter	\<MENU\> \<METER\>
Notch (auto or manual)	\<FUNCTION\> \<NOTCH\>
Notch width (manual)	\<hold NOTCH\> \<MULTI\> \<NOTCH WIDTH\>
Panadapter Fixed Edges	Hold \<M.SCOPE\> Hold \<EXPD/SET\> \<Fixed Edges\>
Playback recorded audio file	\<MENU\> \<RECORD\> \<Play Files\>
Power on splash screen settings	\<MENU\> \<SET\> \<Display\> \<Opening Message\> \<MENU\> \<SET\> \<Display\> \<Power ON Check\>
Record an audio file	\<MENU\> \<RECORD\> \<\<REC Start\>\> \<MENU\> \<RECORD\> \<\<REC Stop\>\>
RF gain / Squelch control	\<MENU\> \<SET\> \<Function\> \<RF/SQL Control\>
Reset (partial or full)	\<MENU\> \<SET\> \<Others\> \<Reset\>
RF Power adjust	\<MULTI\> \<RF Power\>
RIT / XIT control function	\<MENU\> \<SET\> \<Function\> \<Quick RIT/ ⊠TX Clear
RTS / DTR settings	\<MENU\> \<SET\> \<Connectors\> \<USB SEND/Keying\>
RTTY Mark frequency	\<MENU\> \<SET\> \<Function\> \<RTTY Mark Frequency\>
RTTY edit messages	\<MENU\> \<DECODE\> \<TX MEM\> \<EDIT\>
RTTY messages via keypad F1 – F4	\<MENU\> \<SET\> \<Connectors\> \<External Keypad\> \<RTTY\>
RTTY Polarity	\<MENU\> \<SET\> \<Function\> \<RTTY Keying Polarity\>
RTTY Shift	\<MENU\> \<SET\> \<Function\> \<RTTY Shift Width\>

RTTY data to USB port	<MENU> <SET> <Connectors> <USB Serial Function>
Scan settings	<MENU> <SCAN>
Screen saver	<MENU> <SET> <Display> <Screen Saver>
Screen capture	<MENU> <SET> <Function> <Screen Capture [POWER] Switch>
Scope settings	Hold <M.SCOPE> Hold <EXPD/SET>
SD Card	<MENU> <SET> <SD Card>
Spectrum Scope settings	<MENU> <SCOPE> or hold down M.SCOPE
Speech language and speed	<MENU> <SET> <Function> <SPEECH>
Split setting	<MENU> <SET> <Function> <SPLIT>
Squelch / RF gain control	<MENU> <SET> <Function> <RF/SQL Control>
Time set	<MENU> <SET> <Time Set>
Tone controls (RX) Treble	<MENU> <SET> <Tone Control/TBW> <RX> <choose mode> <RX Treble>
Tone controls (RX) Bass	<MENU> <SET> <Tone Control/TBW> <RX> <choose mode> <RX Bass>
Tone controls (TX)	<MENU> <SET> <Tone Control/TBW> <TX>
Transmit bandwidth	<MENU> <SET> <Tone Control/TBW> <TX>
Transmit delay	<MENU> <SET> <Function> <TX Delay>
USB port DTR/RTS settings	<MENU> <SET> <Connectors> <USB SEND/Keying>
USB SEND Keying	<MENU> <SET> <Connectors> <USB SEND/Keying>
VFO tuning step	<MENU> <SET> <Function> <MAIN DIAL Auto TS>
Voice keyer monitor	<MENU> <VOICE> <REC/SET> <SET> <Auto Monitor>
Voice messages	<MENU> <VOICE> <REC/SET> <SET>
Voice messages F1 – F4	<MENU> <SET> <Connectors> <External Keypad> <VOICE>
VOX settings	Hold <VOX> button
Waterfall settings	Hold <M.SCOPE> Hold <EXPD/SET>
WSJT-X Echo setting	<MENU> <SET> <Connectors> <CI-V> <CI-V USB Echo Back> <ON>